Advances in Intelligent Systems and Computing

Volume 321

Series editor

Janusz Kacprzyk, Polish Academy of Sciences, Warsaw, Poland
e-mail: kacprzyk@ibspan.waw.pl

About this Series

The series "Advances in Intelligent Systems and Computing" contains publications on theory, applications, and design methods of Intelligent Systems and Intelligent Computing. Virtually all disciplines such as engineering, natural sciences, computer and information science, ICT, economics, business, e-commerce, environment, healthcare, life science are covered. The list of topics spans all the areas of modern intelligent systems and computing.

The publications within "Advances in Intelligent Systems and Computing" are primarily textbooks and proceedings of important conferences, symposia and congresses. They cover significant recent developments in the field, both of a foundational and applicable character. An important characteristic feature of the series is the short publication time and world-wide distribution. This permits a rapid and broad dissemination of research results.

Advisory Board

Chairman

Nikhil R. Pal, Indian Statistical Institute, Kolkata, India
e-mail: nikhil@isical.ac.in

Members

Rafael Bello, Universidad Central "Marta Abreu" de Las Villas, Santa Clara, Cuba
e-mail: rbellop@uclv.edu.cu

Emilio S. Corchado, University of Salamanca, Salamanca, Spain
e-mail: escorchado@usal.es

Hani Hagras, University of Essex, Colchester, UK
e-mail: hani@essex.ac.uk

László T. Kóczy, Széchenyi István University, Győr, Hungary
e-mail: koczy@sze.hu

Vladik Kreinovich, University of Texas at El Paso, El Paso, USA
e-mail: vladik@utep.edu

Chin-Teng Lin, National Chiao Tung University, Hsinchu, Taiwan
e-mail: ctlin@mail.nctu.edu.tw

Jie Lu, University of Technology, Sydney, Australia
e-mail: Jie.Lu@uts.edu.au

Patricia Melin, Tijuana Institute of Technology, Tijuana, Mexico
e-mail: epmelin@hafsamx.org

Nadia Nedjah, State University of Rio de Janeiro, Rio de Janeiro, Brazil
e-mail: nadia@eng.uerj.br

Ngoc Thanh Nguyen, Wroclaw University of Technology, Wroclaw, Poland
e-mail: Ngoc-Thanh.Nguyen@pwr.edu.pl

Jun Wang, The Chinese University of Hong Kong, Shatin, Hong Kong
e-mail: jwang@mae.cuhk.edu.hk

More information about this series at http://www.springer.com/series/11156

Rajkumar Buyya · Sabu M. Thampi
Editors

Intelligent Distributed Computing

 Springer

Editors
Rajkumar Buyya
Department of Computing and Information
The University of Melbourne
Melbourne
Australia

Sabu M. Thampi
Indian Institute of Information Technology
and Management – Kerala (IIITMK)
Technopark Campus
Kerala
India

ISSN 2194-5357   ISSN 2194-5365   (electronic)
ISBN 978-3-319-11226-8   ISBN 978-3-319-11227-5   (eBook)
DOI 10.1007/978-3-319-11227-5

Library of Congress Control Number: 2014949498

Springer Cham Heidelberg New York Dordrecht London

Printed on acid-free paper

Springer is part of Springer Science+Business Media (www.springer.com)

Preface

The third International Symposium on Intelligent Informatics (ISI-2014) provided a forum for sharing original research results and practical development experiences among experts in the emerging areas of Intelligent Informatics. This edition was co-located with third International Conference on Advances in Computing, Communications and Informatics (ICACCI-2014), September 24–27, 2014. The Symposium was hosted by Galgotias College of Engineering & Technology, Greater Noida, Delhi, India.

ISI-2014 had two tracks: 'Advances in Intelligent Informatics' and a special track on 'Intelligent Distributed Computing'. In response to the call for papers, 134 papers were submitted for the Intelligent Informatics Track and 63 submissions for the Intelligent Distributed Computing track. All the papers were evaluated on the basis of their significance, novelty, and technical quality. Each paper was rigorously reviewed by the members of the program committee. This book contains a selection of refereed and revised papers of Intelligent Distributed Computing Track originally presented at the symposium. In this track, 18 regular papers and 7 short papers were accepted. The peer-reviewed papers selected for this Track cover several Distributed Computing and related topics including Peer-to-Peer Networks, Cloud Computing, Mobile Clouds, Wireless Sensor Networks, and their applications.

Many people helped to make ISI-2014 a successful event. Credit for the quality of the conference proceedings goes first and foremost to the authors. They contributed a great deal of effort and creativity to produce this work, and we are very thankful that they chose ISI-2014 as the place to present it. Thanks to all members of the Technical Program Committee, and the external reviewers, for their hard work in evaluating and discussing papers. We wish to thank all the members of the Steering Committee and Organising Committee, whose work and commitment were invaluable. Our most sincere thanks go to all keynote speakers who shared with us their expertise and knowledge. The EDAS conference system proved very helpful during the submission, review, and editing phases.

We thank the Galgotias College of Engineering & Technology, Greater Noida, Delhi for hosting the conference. Sincere thanks to Suneel Galgotia, Chairman, GEI, Dhruv Galgotia, CEO, GEI, R. Sundaresan, Director, GCET and Bhawna Mallick, Local Arrangements Chair for their valuable suggestions and encouragement.

We wish to express our thanks to Thomas Ditzinger, Senior Editor, Engineering/Applied Sciences Springer-Verlag for his help and cooperation.

September 2014   Rajkumar Buyya
                                                                                  Sabu M. Thampi

Organization

ISI'14 Committee

General Chair

Kuan-Ching Li   Providence University, Taiwan

TPC Chairs

El-Sayed M. El-Alfy	King Fahd University of Petroleum and Minerals, Saudi Arabia
Selwyn Piramuthu	University of Florida, USA
Thomas Hanne	University of Applied Sciences, Switzerland

TPC Members

A.B.M. Moniruzzaman	Daffodil International University, Bangladesh
A.F.M. Sajidul Qadir	Samsung R&D Institute-Bangladesh, Bangladesh
Abdelmajid Khelil	Huawei European Research Center, Germany
Aboul Ella Hassanien	University of Cairo, Egypt
Adel Alimi	REGIM, University of Sfax, National School of Engineers, Tunisia
Afshin Shaabany	University of Fasa, Iran
Agostino Bruzzone	University of Genoa, Italy
Aitha Nagaraju	CURAJ, India
Ajay Jangra	KUK University, Kurukshetra, Haryana, India
Ajay Singh	Multimedia University, Malaysia
Akash Singh	IBM, USA
Akhil Gupta	Shri Mata Vaishno Devi University, India
Akihiro Fujihara	Fukui University of Technology, Japan

Alex James   Nazarbayev University, Kazakhstan
Ali Yavari   KTH Royal Institute of Technology, Sweden
Amit Acharyya   IIT HYDERABAD, India
Amit Gautam   SP College of Engineering, India
Amudha J.   Anrita Vishwa Vidyapeetham, India
Anca Daniela Ionita   University Politehnica of Bucharest,
                                        Romania
Angelo Trotta   University of Bologna, Italy
Angelos Michalas   Technological Education Institute of
                                        Western Macedonia, Greece
Anirban Kundu   Kuang-Chi Institute of Advanced
                                        Technology, P.R. China
Aniruddha Bhattacharjya   Amrita School of Engineering Bangalore,
                                        India
Anjana Gosain   Indraprastha University, India
Antonio LaTorre   Universidad Politécnica de Madrid, Spain
Arpan Kar   Indian Institute of Management, Rohtak,
                                        India
Ash Mohammad Abbas   Aligarh Muslim University, India
Ashish Saini   Dayalbagh Educational Institute, India
Ashraf S.   IITMK, India
Athanasios Pantelous   University of Liverpool, United Kingdom
Atsushi Takeda   Tohoku Gakuin University, Japan
Atul Negi   University of Hyderabad, India
Azian Azamimi Abdullah   Universiti Malaysia Perlis, Malaysia
B.H. Shekar   Mangalore University, India
Belal Abuhaija   University of Tabuk, Saudi Arabia
Bhushan Trivedi   GLS Institute Of Computer Technology,
                                        India
Bilal Gonen   University of West Florida, USA
Bilal Khan   University of Sussex, United Kingdom
Chia-Hung Lai   National Cheng Kung University, Taiwan
Chia-Pang Chen   National Taiwan University, Taiwan
Chien-Fu Cheng   Tamkang University, Taiwan
Chiranjib Sur   ABV-Indian Institute of Information
                                        Technology & Management, Gwalior,
                                        India
Chunming Liu   T-Mobile USA, USA
Ciprian Dobre   University Politehnica of Bucharest,
                                        Romania
Ciza Thomas   Indian Institute of Science, India
Constandinos Mavromoustakis   University of Nicosia, Cyprus
Dalila Chiadmi   Mohammadia School of Engineering,
                                        Morocco

Daniela Castelluccia	University of Bari, Italy
Deepak Mishra	IIST, India
Deepti Mehrotra	AMITY School of Engineering and Technology, India
Demetrios Sampson	University of Piraeus, Greece
Dennis Kergl	Universität der Bundeswehr München, Germany
Dhananjay Singh	Hankuk University of Foreign Studies, Korea
Dimitrios Stratogiannis	National Technical University of Athens, Greece
Durairaj Devaraj	Kalasalingam University, India
Emilio Jiménez Macías	University of La Rioja, Spain
Evgeny Khorov	IITP RAS, Russia
Farrah Wong	Universiti Malaysia Sabah, Malaysia
Fikret Sivrikaya	Technische Universität Berlin, Germany
G. Thakur	MANIT Bhopal, India
Gancho Vachkov	The University of the South Pacific (USP), Fiji
Gorthi Manyam	IIST, India
Gregorio Romero	Universidad Politecnica de Madrid, Spain
Grienggrai Rajchakit	Maejo University, Thailand
Gwo-Jiun Horng	Fortune Institute of Technology, Taiwan
Habib Kammoun	University of Sfax, Tunisia
Habib Louafi	École de Technologies Supérieure (ETS), Canada
Haijun Zhang	Beijing University of Chemical Technology, P.R. China
Hajar Mousannif	Cadi Ayyad University, Morocco
Hammad Mushtaq	University of Management & TEchnology, Pakistan
Hanen Idoudi	National School of Computer Science - University of Manouba, Tunisia
Harikumar Sandhya	Amrita Vishwa Vidyapeetham, India
Hemanta Kalita	North Eastern Hill University, India
Hideaki Iiduka	Kyushu Institute of Technology, Japan
Hossam Zawbaa	Beni-Suef University, Egypt
Hossein Malekmohamadi	University of Lincoln, United Kingdom
Huifang Chen	Zhejiang University, P.R. China
Igor Salkov	Donetsk National University, Ukraine
J. Mailen Kootsey	Simulation Resources, Inc., USA
Jaafar Gaber	UTBM, France
Jagdish Pande	Qualcomm Inc., USA
Janusz Kacprzyk	Polish Academy of Sciences, Poland

Mohamed Moussaoui   Abdelmalek Esaadi UniversitY, Morocco
Mohammad Monirujjaman
   Khan   University of Liberal Arts Bangladesh,
    Bangladesh
Mohammed Mujahid Ulla Faiz   King Fahd University of Petroleum and
    Minerals (KFUPM), Saudi Arabia
Mohand Lagha   Saad Dahlab University of Blida - Blida -
    lgeria, Algeria
Mohd Ramzi Mohd Hussain   International Islamic University Malaysia,
    Malaysia
Monica Chis   Frequentis AG, Romania
Monika Gupta   GGSIPU, India
Mukesh Taneja   Cisco Systems, India
Mustafa Khandwawala   University of North Carolina at Chapel Hill,
    USA
Muthukkaruppan Annamalai   Universiti Teknologi MARA, Malaysia
Naveen Aggarwal   Panjab University, India
Nestor Mora Nuñez   Cadiz University, Spain
Nico Saputro   Southern Illinois University Carbondale,
    USA
Nisheeth Joshi   Banasthali University, India
Noor Mahammad Sk   Indian Institute of Information Technology
    Design and Manufacturing (IIITDM),
    India
Nora Cuppens-Boulahia   IT TELECOM Bretagne, France
Olaf Maennel   Loughborough University, United Kingdom
Omar ElTayeby   Clark Atlanta University, USA
Oskars Ozolins   Riga Technical University, Latvia
Otavio Teixeira   Centro Universitário do Estado do Pará
    (CESUPA), Brazil
Pedro Gonçalves   Universidade de Aveiro, Portugal
Peiyan Yuan   Henan Normal University, P.R. China
Petia Koprinkova-Hristova   Bulgarian Academy of Sciences, Bulgaria
Philip Moore   Lanzhou University, United Kingdom
Praveen Srivastava   Indian Institute of Management (IIM),
    India
Pravin Patil   Graphic Era University Dehradun, India
Qin Lu   University of Technology, Sydney, Australia
Rabeb Mizouni   Khalifa University, UAE
Rachid Anane   Coventry University, United Kingdom
Rafael Pasquini   Federal University of Uberlândia - UFU,
    Brazil
Rajeev Kumaraswamy   Network Systems & Technologies
    Private Ltd., India

Rajeev Shrivastava   MPSIDC, India
Rajib Kar   National Institute of Technology, Durgapur,
                                India
Rakesh Nagaraj   Amrita School of Engineering, India
Rama Garimella   IIIT Hyderabad, India
Ranjan Das   Indian Institute of Technology Ropar, India
Rashid Ali   College of Computers and Information
                                Technology, Taif University,
                                Saudi Arabia
Raveendranathan Kl C   University of Kerala, India
Ravibabu Mulaveesala   Indian Institute of Technology Ropar, India
Rivindu Perera   Auckland University of Technology,
                                New Zealand
Rubita Sudirman   Universiti Teknologi Malaysia, Malaysia
Ryosuke Ando   Toyota Transportation Research Institute
                                (TTRI), Japan
Sakthi Balan   Infosys Ltd., India
Salvatore Venticinque   Second University of Naples, Italy
Sameer Saheerudeen
    Mohammed   National Institute of Technology Calicut,
                                India
Sami Habib   Kuwait University, Kuwait
Sasanko Gantayat   GMR Institute of Technology, India
Satish Chandra   Jaypee Institute of Information Technology,
                                India
Satya Ghrera   Jaypee University of Information
                                Technology, India
Scott Turner   University of Northampton,
                                United Kingdom
Selvamani K.   Anna University, India
Shalini Batra   Thapar University, India
Shanmugapriya D.   Avinashilingam Institute, India
Sheeba Rani   IIST Trivandrum, India
Sheng-Shih Wang   Minghsin University of Science and
                                Technology, Taiwan
Shubhajit Roy Chowdhury   IIIT Hyderabad, India
Shuping Liu   University of Southern California, USA
Shyan Ming Yuan   National Chiao Tung University, Taiwan
Siby Abraham   University of Mumbai, India
Simon Fong   University of Macau, Macao
Sotiris Karachontzitis   University of Patras, Greece
Sotiris Kotsiantis   University of Patras, Greece
Sowmya Kamath S.   National Institute of Technology, Surathkal,
                                India

Sriparna Saha	IIT Patna, India
Su Fong Chien	MIMOS Berhad, Malaysia
Sujit Mandal	National Institute of Technology, Durgapur, India
Suma V.	Dayananda Sagar College of Engineering, VTU, India
Suryakanth Gangashetty	IIIT Hyderabad, India
Tae (Tom) Oh	Rochester Institute of Technology, USA
Teruaki Ito	University of Tokushima, Japan
Tilokchan Irengbam	Manipur University, India
Traian Rebedea	University Politehnica of Bucharest, Romania
Tutut Herawan	Universiti Malaysia Pahang, Malaysia
Usha Banerjee	College of Engineering Roorkee, India
V. Vityanathan	SASTRA Universisty, India
Vatsavayi Valli Kumari	Andhra University, India
Veronica Moertini	Parahyangan Catholic University, Bandung, Indonesia
Vikrant Bhateja	Shri Ramswaroop Memorial Group of Professional Colleges, Lucknow (UP), India
Visvasuresh Victor Govindaswamy	Concordia University, USA
Vivek Sehgal	Jaypee University of Information Technology, India
Vivek Singh	Banaras Hindu University, India
Wan Hussain Wan Ishak	Universiti Utara Malaysia, Malaysia
Wei Wei	Xi'an University of of Technology, P.R. China
Xiaoya Hu	Huazhong University of Science and Technology, P.R. China
Xiao-Yang Liu	Columbia University, P.R. China
Yingyuan Xiao	Tianjin University of Technology, P.R. China
Yong Liao	NARUS INC., USA
Yoshitaka Kameya	Meijo University, Japan
Yuming Zhou	Nanjing University, P.R. China
Yu-N Cheah	Universiti Sains Malaysia, Malaysia
Zhenzhen Ye	IBM, USA
Zhijie Shen	Hortonworks, Inc., USA
Zhuo Lu	Intelligent Automation, Inc., USA

Additional Reviewers

Abdelhamid Helali	ISIMM, Tunisia
Adesh Kumar	UPES, India
Adriano Prates	Universidade Federal Fluminense, Brazil
Amitesh Rajput	Sagar Institute of Science & Technology, Bhopal, India
Antonio Cimmino	Zurich University of Applied Sciences, Switzerland
Aroua Hedhili	SOIE, National School of Computer Studies (ENSI), Tunisia
Azhana Ahmad	Universiti Tenaga Nasional, Malaysia
Behnam Salimi	UTM, Malaysia
Carolina Zato	University of Salamanca, Spain
Damandeep Kaur	Thapar University, India
Devi Arockia Vanitha	The Standard Fireworks Rajaratnam College for Women, Sivakasi & India
Divya Upadhyay	Amity School of Engineering & Technology, Amity University Noida, India
Gopal Chaudhary	NSIT, India
Hanish Aggarwal	Indian Institute of Technology Roorkee, India
Indrajit De	MCKV Institute of Engineering, India
Iti Mathur	Banasthali University, India
Kamaldeep Kaur	Guru Gobind Singh Indraprastha University, India
Kamyar Mehranzamir	UTM, Malaysia
Kotaiah Bonthu	Babasaheb Bhimrao Ambedkar University, Lucknow, India
Laila Benhlima	Mohammed V-Agdal University, Mohammadia School of Engineering, Morocco
M. Karuppasamypandiyan	Kalasalingam University, India
Mahalingam Pr.	Muthoot Institute of Technology and Science, India
Manisha Bhende	University of Pune, India
Mariam Kassim	Anna University, India
Mohammad Abuhweidi	University of Malaya, Malaysia
Mohammad Hasanzadeh	Amirkabir University of Technology, Iran
Muhammad Imran Khan	University of Toulouse, France
Muhammad Murtaza	University of Engineering and Technology Lahore, Pakistan
Muhammad Rafi	FAST-NU, Pakistan
Mukundhan Srinivasan	Indian Institute of Science, India
Naresh Kumar	GGSIPU, India

Nasimi Eldarov	Saarland University, Germany
Nattee Pinthong	Mahanakorn University of Technology, Thailand
Nouman Rao	Higher Education Commission of Pakistan, Pakistan
Omar Al saif	Mosul University, Iraq
Orlewilson Maia	Federal University of Minas Gerais, Brazil
Pankaj Kulkarni	Rajasthan Technical University, India
Paulus Sheetekela	MIPT SU, Russia
Pooja Tripathi	IPEC, India
Prakasha Shivanna	RNS Institute of Technology, India
Pratiyush Guleria	Himachal Pradesh University Shimla, India
Preetvanti Singh	Dayal Bagh Educational Institute, India
Prema Nedungadi	Amrita University, India
Rathnakar Achary	Alliance Business Academy, India
Rohit Thanki	C U Shah Unversity, India
Roozbeh Zarei	Victoria University, Australia
Saida Maaroufi	Ecole Polytechnique de Montréal, Canada
Saraswathy Shamini Gunasekaran	Universiti Tenaga Nasional, Malaysia
Sarvesh Sharma	BITS-Pilani, India
Senthil Sivakumar	St. Joseph University, Tanzania
Seyedmostafa Safavi	Universiti Kebangsaan Malaysia, Malaysia
Shanmuga Sundaram Thangavelu	Amrita Vishwa Vidyapeetham, India
Shruti Kohli	Birla institute of Technology, India
Sudhir Rupanagudi	WorldServe Education, India
Thenmozhi Periasamy	Avinashilingam University, India
Tripty Singh	Amrita Vishwa Vidyapeetham, India
Umer Abbasi	Universiti Teknologi PETRONAS, Malaysia
Vaidehi Nedu	Dayananda Sagar College of Engineering, India
Vijender Solanki	Anna University, Chennai, India
Vinita Mathur	JECRC, Jaipur, India
Vipul Dabhi	Information Technology Department, Dharmsinh Desai University, India
Vishal Gupta	BITS, India
Vrushali Kulkarni	College of Engineering, Pune, India
Yatendra Sahu	Samrat Ashok Technological Institute, India
Yogesh Meena	Hindustan Institute of Technology and Management, India
Yogita Thakran	Indian Institute of Technology Roorkee, India
Zhiyi Shao	Shaanxi Normal University, P.R. China

Steering Committee

Antonio Puliafito	MDSLab - University of Messina, Italy
Axel Sikora	University of Applied Sciences Offenburg, Germany
Bharat Bhargava	Purdue University, USA
Chandrasekaran K.	NITK, India
Deepak Garg, Chair	IEEE Computer Society Chapter, IEEE India Council
Dilip Krishnaswamy	IBM Research - India
Douglas Comer	Purdue University, USA
El-Sayed M. El-Alfy	King Fahd University of Petroleum and Minerals, Saudi Arabia
Gregorio Martinez Perez	University of Murcia, Spain
Hideyuki Takagi	Kyushu University, Japan
Jaime Lloret Mauri	Polytechnic University of Valencia, Spain
Jianwei Huang	The Chinese University of Hong Kong, Hong Kong
John F. Buford	Avaya Labs Research, USA
Manish Parashar	Rutgers, The State University of New Jersey, USA
Mario Koeppen	Kyushu Institute of Technology, Japan
Nallanathan Arumugam	King's College London, United Kingdom
Nikhil R. Pal	Indian Statistical Institute, Kolkata, India
Pascal Lorenz	University of Haute Alsace, France
Raghuram Krishnapuram	IBM Research - India
Raj Kumar Buyya	University of Melbourne, Australia
Sabu M. Thampi	IIITM-K, India
Selwyn Piramuthu	University of Florida, USA
Suash Deb, President	Intl. Neural Network Society (INNS), India Regional Chapter

ICACCI Organising Committee

Chief Patron

Suneel Galgotia, Chairman	GEI, Greater Noida

Patrons

Dhruv Galgotia, CEO	GEI, Greater Noida
R. Sundaresan, Director	GCET, Greater Noida

General Chairs

Sabu M. Thampi   IIITM-K, India
Demetrios G. Sampson   University of Piraeus, Greece
Ajith Abraham   MIR Labs, USA & Chair, IEEE SMCS TC
                                  on Soft Computing

Program Chairs

Peter Mueller   IBM Zurich Research Laboratory,
                                    Switzerland
Juan Manuel Corchado
  Rodriguez   University of Salamanca, Spain
Javier Aguiar   University of Valladolid, Spain

Industry Track Chair

Dilip Krishnaswamy   IBM Research Labs, Bangalore, India

Workshop and Symposium Chairs

Axel Sikora   University of Applied Sciences Offenburg,
                                    Germany
Farag Azzedin   King Fahd University of Petroleum and
                                    Minerals, Saudi Arabia
Sudip Misra   Indian Institute of Technology, Kharagpur,
                                    India

Special Track Chairs

Amit Kumar   BioAxis DNA Research Centre, India
Debasis Giri   Haldia Institute of Technology, India

Demo/Posters Track Chair

Robin Doss   School of Information Technology,
                                    Deakin University, Australia

Keynote/Industry Speakers Chairs

Al-Sakib Khan Pathan   IIUM, Malaysia
Ashutosh Saxena   Infosys Labs, India
Shyam Diwakar   Amrita Vishwa Vidyapeetham Kollam,
                                    India

Tutorial Chairs

Sougata Mukherjea   IBM Research-India
Praveen Gauravaram   Tata Consultancy Services Ltd., Hyderabad,
                                    India

Doctoral Symposium Chairs

Soura Dasgupta   The University of Iowa, USA
Abdul Quaiyum Ansari   Dept. of Electrical Engg., Jamia Millia
                                  Islamia, India
Praveen Ranjan Srivastava   Indian Institute of Management (IIM),
                                  Rohtak, India

Organizing Chair

Bhawna Mallick   Galgotias College of Engineering &
                                  Technology (GCET), India

Organizing Secretaries

Sandeep Saxena   Dept. of CSE, GCET
Rudra Pratap Ojha   Dept. of IT, GCET

Publicity Chairs

Lucknesh Kumar   Dept. of CSE, GCET
Dharm Raj   Dept. of IT, GCET

Contents

Contents   XXI

Customization of Recommendation System Using Collaborative Filtering Algorithm on Cloud Using Mahout

Thangavel Senthil Kumar and Swati Pandey

Abstract. Recommendation System helps people in decision making regarding an item/person. Growth of World Wide Web and E-commerce are the catalyst for recommendation system. Due to large size of data, recommendation system suffers from scalability problem. Hadoop is one of the solutions for this problem. Collaborative filtering is a machine learning algorithm and Mahout is an open source java library which favors collaborative filtering on Hadoop environment. The paper discusses on how recommendation system using collaborative filtering is possible using Mahout environment. The performance of the approach has been presented using Speedup and efficiency.

1   Introduction

In fast growing world, since time is on its heel, people do not want to go shop by shop and buy the best item according to their requirement. To save time, everyone wants to buy things in home in reasonable cost. They prefer online shopping, online suggestion for an item, so that they can take a decision on a particular item which may be suitable for a particular. In such scenario Recommendation System plays a vital role. When the questions are "Whether I will like this item", "I want to buy an item of a particular type which suits according to my taste", "Can you suggest me an item which we may like?", the feasible answers can be obtained through recommendation system. It helps in recommending items of a similar type as well as predicting an item, whether it will be liked by user or not.

Thangavel Senthil Kumar · Swati Pandey
Computer Science and Engineering Department, Amrita Vishwa Vidyapeetham,
Coimbatore
e-mail: t_senthilkumar@cb.amrita.edu, swati.padeycs@gmail.com

© Springer International Publishing Switzerland 2015
R. Buyya and S.M. Thampi (eds.), *Intelligent Distributed Computing*,
Advances in Intelligent Systems and Computing 321, DOI: 10.1007/978-3-319-11227-5_1

1

For recommendation, our proposed system uses collaborative filtering machine learning algorithm. Collaborative filtering (CF) is a machine learning algorithm which is widely used for recommendation purpose. Collaborative filtering finds nearest neighbor based on the similarities. The metric of collaborative filtering is the rating given by the user on a particular item.

Different users give different ratings to items. Users, who give almost same rating to items, are the nearest neighbors. In case of User based collaborative filtering, based on the ratings given by the users, nearest neighbors has been find. Item based collaborative filtering predicts the similarity among items. To recommend an item, items which are liked by the user in his past have been found. Item which is similar to those items has been recommended [10].

Internet contains a huge volume of data for recommendation purpose. Due to size of data, if recommendation computation has been done in single system, then performance may degrade, and we cannot find an efficient solution. Hence we require distributed environment so that computation can be increased and performance of recommendation system gets improve. Cloud is a Distributed System. An open source cloud environment Hadoop provides distributed environment [21]. Due to Map-Reduce programming, it provides result efficiently and effectively in less amount of time. Proposed system has been modeled on Hadoop. Mahout is an open source java library which favors Collaborative Filtering. Mahout favors Hadoop for recommendation [13]. Cloud computing relies on sharing of resources to achieve coherence and economies of scale, similar to a utility (like the electricity grid) over a network. At the foundation of cloud computing is the broader concept of converged infrastructure and shared services. Cloud has emerged as a new face of fast and efficient computing. It provides a wide variety of benefits like scalability of computing power, global accessibility of data and services, efficient storage of data etc. the cloud has extended its reach to many sectors and is intensively used in sectors with necessity for high performance computing, one such sector is entertainment. The term "cloud computing" is mostly used to sell hosted services in the sense of application service provisioning that run client server software at a remote location. Such services are given popular acronyms like 'SaaS' (Software as a Service), 'PaaS' (Platform as a Service), 'IaaS' (Infrastructure as a Service), 'HaaS' (Hardware as a Service) and finally 'EaaS' (Everything as a Service). End users access cloud-based applications through a web browser, thin client or mobile app while the business software and user's data are stored on servers at a remote location. Examples include Amazon Web Services and Google App engine, which allocate space for a user to deploy and manage software "in the cloud". This project utilizes two services provided by cloud namely platform as a service (PaaS) and infrastructure as a service (IaaS).

Rest of the paper has been arranged in different sections. Section 2 briefly describe about related work that has been done. Section 3 elaborates the proposed model for Recommendation System. Section 4 describes data set and results. Section 5 focuses on conclusion and future enhancement.

2   Related Work

Zhi-Dan Zhao and Ming-Sheng Shang [12] have used based Collaborative Filtering using Hadoop as distributed framework. The approach is scalable but the response time taken for a single user could not be reduced. Carlos E. Seminario a David C. Wilson [4] use Mahout for recommendation. Use of mahout for Collaborative Filtering has enhanced the accuracy.

Xiao Yan Shi, Hong Wu Ye and Song Jie Gong [27] integrated user based and item based collaborative filtering algorithm to increase performance. Manos Papagelis and Dimitris Plexousakis [8] did qualitative analysis on user based and item based collaborative filtering with implicit and explicit ratings. According to their analysis, prediction based on explicit rating is better than implicit rating. Trouong Khanh Quan, Ishikawa Fuyuki, Honiden Shinichi [24] clustered the items on stability of user similarity and applied Collaborative Filtering. Final clustering of the items may be locally optimal but the prediction accuracy has been increased.

Yanhong Guo, Xuefen Cheng, Dahai Dong, Chunyu Luo and Rishuang Wang [26] used CF based on trust factor. For calculating similarities Cosine correlation and Pearson   correlation has been used. They found that CF based on trust factor is better than traditional CF. Mustansar Ali Ghazanfar and Adam Prugel Bnnett [20] had built hybrid recommendation system which combines the rating, feature and demographic information of item. Yajie Hu, Ziqi Wang, Wei Wu, Jianzhong Guo and Ming Zhang [25] used Semantic distance measurement and considered the features of movie. They are able to give a list of recommended movie along with stars of that movie. Dhoha Almazro, Ghadeer Shahatah, Lamia Albbulkarim, Mona Kherees, Romy Martinez and William Nzoukou [17] collected Demographic information of user, clustered all items and then combined both user-based and item-based. It could be taken for medical and road accident datasets [22][23].

Hee Choon Lee, Seok Jun Lee, Young Jun Chung [18] did a study on improved Collaborative filtering algorithm. They have used Neighborhood CF (NBCFA), Correspondence mean algorithm (CMA). They found that the preference prediction performance of CMA is better than NBCFA. Kai Yu, Xiaowei Xu, Jianhua Tao, Martin Aster and EansPeter Kriegel [19] proposed instance selection techniques for memory based collaborative filtering. It reduces the storage requirement of training data, but there is scalability problem for big data. Alexandros Karatzoglou, Alex Smola and Markus Weimer [14] proposed CF with hashing using It-intensive loss function, Huber loss function. This model can be scaled to bigger data-set on large server and to still data-set on small machines, but time complexity is the problem. Aristomenis S. Lampropoulos and George A. Tsihrintzis [2] did survey on recommendation machine learning algorithms. Sarwar B. et al. [10], Mukund Deshpande et al. [9] and Badrul Sarwar et al. [15] worked on item based collaborative filtering for recommendation.

3   Proposed System

This paper is focusing to customize a recommendation system using Collaborative Filtering (CF) and clustering techniques. For our approach, we use Apache Hadoop (a widely used open source platform which implements Map-Reduce programming model) and Apache Mahout (An Open source library of scalable data mining algorithms, where it forms a core of the distributed recommender module).

   The proposed algorithm is worked in two parts. In first phase, we obtain the results for recommendation by applying User-based CF and Item-based CF separately. In second phase, we combine the results obtain from user-based CF and item-based CF.

3.1   *User-Based Collaborative Filtering*

The dataset is firstly loaded into Hadoop distributed file system (HDFS). Then we perform User-based CF using Mahout. We take rating matrix, in which each row represents user and column represents item, corresponding row-column value represents rating which is given by a user to an item. Absence of rating value indicates that user has not rated the item yet. There are many similarity measurement methods to compute nearest neighbors. We have used Pearson correlation coefficient to find similarity between two users. Hadoop is used to calculate the similarity. The output of the Hadoop Map phase i.e. userid and corresponding itemid are passed to reduce phase. In reduce phase, output has been generated and sorted according to userid. Output again has been stored in HDFS. The architecture diagram for User-based CF can be shown in fig. 1.

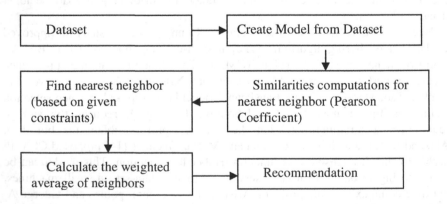

Fig. 1 User-based Collaborative Filtering

3.2   Item-Based Collaborative Filtering

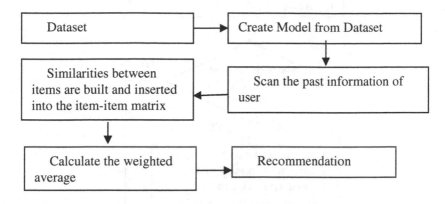

Fig. 2 Item-based Collaborative Filtering

Dataset is loaded into HDFS, then using Mahout we performs Item-based CF. Past information of the user, i.e. the ratings they gave to items are collected. With the help of this information the similarities between items are built and inserted into item to item matrix. Algorithm selects items which are most similar to the items rated by the user in past. In next step, based on top-N recommendation, target items are selected. The processing of Item-based CF can be described through fig. 2.

3.3   Combined Result

In case of user-based CF, if nearest neighbors (similar users) are not in enough number, i.e. taste of target user is not similar to many users then the recommending an item for that particular user may be not accurate. Item-based CF is based on the past information of the user, so it works well in such cases. The user-based and item-based results that are stored in HDFS are taken and then we combine these results based on threshold value. Suppose the threshold by increasing threshold value, probability of recommending correct item get increases because it has been liked by many users. Flow can be seen in fig. 3.

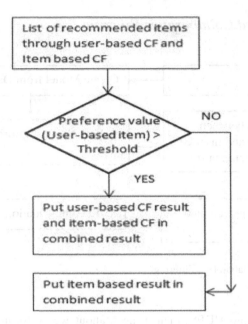

Fig. 3 Combined Result

4   Experiment and Result

4.1   Dataset

For experiment, we have used MovieLens dataset of size 1M. The dataset contains 10000054 ratings and 95580 tags applied to 10681 movies by 71567 users. There are three files, movies.dat, ratings.dat and tags.dat [24]. Ratings data file has atleast three columns; those are UserId, given by user to movie [13].

4.2   Result Analysis

For movie recommendation, important factor is the list of recommended items as soon as possible. Since we are using Hadoop, speedup and efficiency varies as number of nodes varies. To analyze this we have obtained the number of movies which are recommended as threshold changes, Speedup and efficiency according to number of nodes.

4.3   Movies vs. Threshold

Threshold value indicates minimum number of users who like an item which has to be recommended. It means when we increase threshold, accuracy in recommending an item increases because more number of users like that item.

When the threshold value increases, the number of items recommended by the users will be reduced. Even though that value is too small, it is the relevant one. Hence such items can be readily recommended to users without any further processes. Comparison graph based on threshold value is given by Fig. 4.

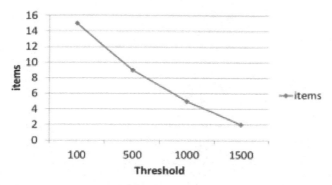

Fig. 4 Comparison graph based on threshold

4.4 *SpeedUp*

Speedup is the measure of the performance. It is defined as the ratio between sequential execution time and the parallel execution time.

$$Speed\ Up = T(1) / T(b)$$

Where T(1) states the execution time taken by single processor. T(b) is the execution time taken by 'b' no. of processor. While running algorithm in Hadoop framework, speedup varies as numbers of nodes vary. As we increase no of nodes, speedup increases. It can be seen in fig. 5.

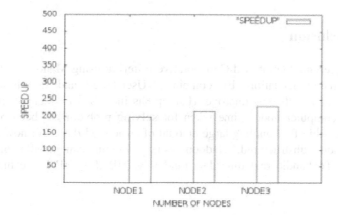

Fig. 5 SpeedUp with respect to no. of nodes

4.5   *Efficiency*

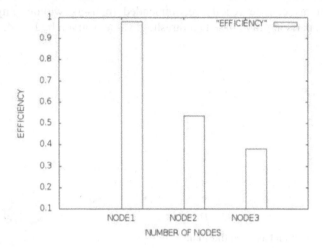

Fig. 6 Efficiency with respect to no. of nodes

It is denoted as the usage of the computational resources. This is the ration between the speedup and no. of processors.

$$\text{Efficiency Ratio} = T(1) \, / \, b.T \, (b)$$

From the above graph (fig. 6) it is visible that if the number of processors is one, the efficiency increases because that processor is fully utilized for the particular program. When the processors increases the efficiency decreases, that means we can utilize the processors for different purposes. This is the one main advantage of the distributed system.

5   Conclusion

This paper presented combined Collaborative Filtering using Mahout on Hadoop for movie recommendation. By combining User-based and Item-based CF, accuracy of the results gets improve. Hadoop has increased throughput. Because of multiple computer nodes, time taken for solving problem has been reduced. Mahout is feasible for handling large amount of structured data. But now a day's data are more unstructured. Hadoop using Mahout can handle big data statistically. To handle real time data randomly, HBASE, HIVE are the better solutions.

References

[1] Lam, C.: Hadoop in Action. Manning Publication (2010)

[2] Lampropoulos, A.S., Tsihrintzis, G.A.: A survey approach to designing recommendation system. Springer (2013)

[3] Tanenbaum, A.S., van Steen, M.: Distributed Systems: Principle and Paradigms, 2nd edn. Pearson Prentice Hall (May 2005)

[4] Seminario, C.E., Wilson, D.C.: Case study evaluation of mahout as a recommender plateform. Presented in Workshop on Recommendation Utility Evaluation: Beyond RMSE, held in Conjunction with ACM in Ireland (2012)

[5] Dean, J., Ghemawat, S.: Map-Reduce: Simplified data processing on large clusters. To appear in OSDI 2004 (2004)

[6] Schafers, J.B., Konstan, J., Riedi, J.: Recommendation Systems in e-commerce. In: 1st ACM Conference on Electronic Commerce, pp. 158–166. ACM Press (1999)

[7] Terveen, L., Hill, W.: Beyond recommender systems: Helping people help each other. In: HCI in The New Millennium. Jack Aarrdl, Addison Wesley (2001)

[8] Papagelis, M., Plexousakis, D.: Qualitative analysis of user based and item based prediction algorithms for recommendation agents. Science Direct 2005 (2005)

[9] Deshpande, M., Karypis, G.: Item based top-N recommendation algorithm. ACM Transactions Information Systems 22(1), 143–177 (2004)

[10] Sarwar, B., Karypis: Item based collaborative filtering algorithms. In: 10th International World Wide Web conference, pp. 285–295 (2001)

[11] Yueping, W., Jianguo, Z.: A research of recommendation algorithm based on cloud model. IEEE (2010)

[12] Zhao, Z.-D., Shang, M.-S.: User based collaborative filtering recommendation algorithm a hadoop. IEEE (2012)

[13] Ajay Reddy, P.K., Chidambaram, M., Anurag, D.V., Karthik, S., Ravi Teja, K., Sri Harish, N., Senthil Kumar, T.: Video Recommender In Open/Closed Systems. International Journal of Research in Engineering and Technology 3 (2014)

[14] Karatzoglou, A., Smola, A., Weimer, M.: Collaborative filtering on a budget. Appear in Proceedings of the 13th International Conference on Artificial Intelligence and Statistics 2010, Italy (2010)

[15] Sarwar, B., Karypis, G., Konstan, J., Riedl, J.: Item based collaborative filtering recommendation algorithms. In: WWW, Hong Kong. ACM (2001)

[16] Tapucu, D., Kasap, S., Tekbacak, F.: Performance comparison of combined collaborative filtering algorithm for Recommender system. IEEE (2012)

[17] Almazro, D., Shahatah, G., Albbulkarim, L., Kherees, M., Martinez, R., Nzoukou, W.: A survey paper on recommendation system. ACM (2010)

[18] Lee, H.C., Lee, S.J., Chung, Y.J.: A study on improved collaborative filtering algorithm for recommendation system. IEEE (2007)

[19] Yu, K., Xu, X., Tao, J., Aster, M., Kriegel, E.-P.: Instance selection techniques for memory based collaborative filtering. SIAM

[20] Ghazanfar, M.A., Bnnett, A.P.: A scalable, accurate hybrid recommender system. In: Third International Conference on Knowledge Discovery and Data Mining, pp. 94–98 (2010)

[21] Senthil Kumar, T., Neetha Susan, T., Johnpaul, C.I.: Performance Analysis of Various Recommendation Algorithms Using Apache Hadoop and Mahout. International Journal of Scientific & Engineering Research 4, 279–287 (2013)

[22] Senthil Kumar, T., Suresh, A., Pai, K.K., Chinnaswamy, P.: Survey on Predictive medical data analysis. International Journal of Engineering Research & Technology 3 (2014)

[23] Senthil Kumar, T., Vishak, J., Sanjeev, S., Sneha, B.: Cloud Based Framework for Road Accident Analysis. International Journal of Computer Science and Mobile Computing 3, 1025–1032 (2014)

[24] Quan, T.K., Fuyuki, I., Shinichi, H.: Improving accuracy of recommendation system by clustering item based on stability of user similarity. In: International Conference on Intelligent Agents, Web Technologies and Internet Commerce (2006)

[25] Shi, X.Y., Ye, H.W., Gong, S.J.: A personalized recommender integrating item based and user based collaborative filtering. IEEE (2008)

[26] Hu, Y., Wang, Z., Wu, W., Guo, J., Zhang, M.: Recommendation for movies and stars using YOGA and IMDB. IEEE (2010)

[27] Guo, Y., Cheng, X., Dong, D., Luo, C., Wang, R.: An improved collaborative filtering algorithm based on trust in e-commerce recommendation system. IEEE (2010)

A Result Verification Scheme for MapReduce Having Untrusted Participants

Gaurav Pareek, Chetanya Goyal, and Mukesh Nayal

Abstract. MapReduce framework is a widely accepted solution for performing data intensive computations efficiently. The master node prepares the input to be distributed among multiple mappers which distribute the reduced task to the reducers. Reducers perform identical set of computations on the reduced data independently. If any one of the reducers works maliciously and does not produce results as desired by the end-user, a significant error in the final output can be observed. Many other distributed computing platforms also face the same problem due to the malicious participants. The problem for MapReduce must be solved keeping into account the data intensive nature of the computations carried out by MapReduce. MapReduce does not provide any mechanism to detect such Lazy Cheating Attacks by a computation provider. In this paper, we propose a generalized defense to this type of attack on statistical computations. The solution does not involve redundant computations on the data to prove the worker malicious. Implementation results on Hadoop show the detection rate of such cheating behavior by the proposed scheme. The accompanying theoretical analysis proves that the solution does not noticeably affect the timeliness and accuracy of the original service.

1 Introduction

MapReduce[1] has become increasingly popular as a parallel data processing model. The framework is designed for performing data intensive computations efficiently. Hadoop proposed by Yahoo as well as thousands of MapReduce applications like Hive, HBase etc. have adopted MapReduce as a parallel processing framework. MapReduce framework is designed to work in distributed computing environments. The framework consists of a master for relegating the tasks to various workers (mappers and reducers). The task is a file residing on the distributed file system (DFS)

Gaurav Pareek · Chetanya Goyal · Mukesh Nayal
BT Kumaon Institute of Technology, Dwarahat, India
e-mail: gauravpareek@curaj.ac.in,
{chetanya9292,mukesh.nayal992}@gmail.com

© Springer International Publishing Switzerland 2015   11
R. Buyya and S.M. Thampi (eds.), *Intelligent Distributed Computing*,
Advances in Intelligent Systems and Computing 321, DOI: 10.1007/978-3-319-11227-5_2

that provides a distributed storage for data in MapReduce. The master assigns a map task (residing on DFS) to a subset of its workers called mappers. The mappers compute the intermediate results and store in the local storage. The mappers then divide the task into partitions $P_1, P_2, ... P_n$ and deliver each partition to reducer in the system. The reducers are notified of the completion of this "Map"phase. When notified, the reducer copies the intermediate result from the mapper to its local storage and performs computations. The results of final computations are written by each reducer to the DFS. The results of computations by each reducer are now combined and the final output is also written to the DFS. This framework, like many distributed computing platforms, has service integrity problems [2-8]. That is, if the computations are not performed accurately by workers the final output can have considerable variation from the actual. Various authors addressed the service integrity problem in different ways. We classify them as Replication-Based, Sampling-Based and Checkpointing-Based Service Integrity Schemes. Some others proposed integrity assurance frameworks for MapReduce. Detailed description of these schemes is given in the section 2. In this paper, we propose a cheating detection scheme that focuses on detecting and blacklisting the malicious mapper/reducer working to accomplish a set of statistical computations. The proposed scheme is a hybrid of sampling-based and checkpointing-based schemes. The proposed scheme is an effective solution to even more powerful attacker model capabilities than the one assumed in this paper. The experiments conducted in the paper are implemented and tested on Hadoop [13], an open source framework for MapReduce. The paper is organised as: the section 2 presents a survey of state-of-the-art of the result verification schemes for various computing platforms including MapReduce, Distributed Computing platforms, Peer-to-peer grids etc. The assumptions of the system and attacker capabilities are listed and discussed in section 3. Section 4 consists the description of the proposed scheme for result verification. In section 5 the statistical display of the amount of threat due to attacks is presented. Also the results showing the detection rate and the discussion on the same is also presented in section 5.

2   Related Work

The need for assuring service integrity arises when there are untrusted participants for performing the part of the overall computation. As a result, this problem has been addressed for various computing models involving untrusted "workers"like Peer-to-Peer, Grid and Volunteer Computing. We classify the approaches for service integrity assurance as Redundant Computations based, Sampling-based and Partitioning/Check-pointing based. In Redundant Computation based or Replication-based schemes the basic idea is to perform the same set of computations by trusted and untrusted infrastructure, then comparing the two results and declaring the untrusted participant as malicious or honest based on a mismatch or match between the two results respectively. In Sampling-Based integrity assurance, the input to the participants is sampled with values in domain whose image is known for a given function. The individual output from every participant can now be cross-checked against the one stored as image a priori the injection of sample to the input. The function to be performed is partitioned into a number of temporal segments in Partitioning/Check-pointing based schemes. The function is a series of computations to be performed and the checkpoints made help determining the possibility of integrity violation at the level of a checkpoint. SecureMR proposed in [9] is an integrity assurance framework for MapReduce. This framework exploits the amenity of MapReduce framework to replication based techniques. In addition to decentralizing integrity verification mechanism using replication-based methods, the framework also modifies the communication protocol so as to provide MapReduce resilience to Replay and Denial of Service (DoS) attacks. Authors in [14] propose a scheme for result verification for untrusted participants. They target the inverted-index computations and propose the watermark injection scheme for detecting the malicious behavior of any mapper(s) or reducer(s). The strength of the defense scheme in [14] lies in the sampling methodology adopted by the authors for injecting the watermark. The paper [6] proposes magic numbers and ringers for making the distributed computations uncheatable. The authors focus on accomplishing the task of computing the key for DES using various participants and making payment according to the amount of work done by the participant. The work in [11] proposes extended ringers as the extension to the ringers and magic numbers in [6]. Once the malicious participant knows of the injection of ringers or any strategic input, the participant may circumvent the effect of these ringers. So, the extended ringers proposed in [11] have additional capability of not making the participants aware of the injected ringers. The scheme employing extended ringers is analysed for both sequential and non-sequential statistical computations like Monte-Carlo simulations, optimizing problems etc. Many papers proposed secure verification schemes for various other computation architectures. Vulnerability to tampering in mobile agents by malicious hosts was first studied in [8] that focuses on studying the need for tamper-proof program execution on untrusted mobile hosts. The basic idea in [8] is to encrypt both data and function to be executed. Airavat [7] is a trust assurance framework for MapReduce that protects privacy and ensures security of computations. Most of all above mentioned result verification schemes are detect the malicious participant using redundant computations either for the whole or part of computation. MapReduce framework has an inherent amenity to redundant computations [9] but it can be resource inefficient to execute same set of computations by a *secondary* computing unit. To match the speed of result delivery, [12] uses a separate dedicated MapReduce computing unit for performing redundant set of tasks. In such cases, the MapReduce computations performed by the separate MapReduce computing unit are specific to the computations submitted to the main MapReduce cluster. Since the problem of cheating in MapReduce arises from the possibility of computation provider being malicious, the computation-specific secondary MapReduce cluster may not be a good idea. So, the proposed scheme not only circumvents the need of submitting the same job to the secondary cluster, but also enables the secondary cluster to be reused by other computations.

3 System and Attacker Model Assumptions

As stated earlier, Hadoop programming model is used for experimental verification and analysis of the proposed scheme. The Hadoop Distributed File System (HDFS) is responsible for storing the input and output of the overall computations. The Combiner function is a part of "Reduce"phase in Hadoop that does the job of combining the individual output of all the reducers.

The main job to be performed by the MapReduce cluster in our case can be formulated as a series of functions $f_1(f_2....(f_n(x)))$. That is, the output of the function applied initially acts as an input to the next function in the series. This way one can easily observe that for computing the standard deviation through MapReduce, the task will be accomplished in two steps: 1) average calculation and 2) using the average to calculate standard deviation. Further, the average calculation requires that the addition of the input numbers be available. The standard deviation requires a constant (mean) to be subtracted from the input data as a constituent step. So, a series of Map-Reduce-Combine phases have to be encountered during the statistical computations. The attacker model capabilities assumed for the paper are as follows:

- The malicious worker wishes to gain through saving computational resources by not evaluating actual set of computations but by evaluating a less computationally expensive function over the input data.
- The malicious worker does not wish to gain through intentionally disturbing accuracy level of the output. In fact, the reducer keeps track of the "kind"of computations being performed so that it is not possible to detect malicious behaviour by merely looking at the output.

The first assumption is to do with the actual intent of the malicious worker which is to save resources. For accomplishing this, the worker evaluates a function g instead of the actual function f at any stage of the overall computations such that $K'(g(x)) = O(K(f(x))$. K and K' being the time/space complexity of the actual and "less-costly"function respectively. Second assumption requires the attackers to be aware of the "kind"of computations being performed and produce results that do not raise a suspicion of attack by looking at the output. For this, the attacker may choose not to behave maliciously during each and every Map-Reduce-Combine cycle. The worker may behave maliciously on a random basis. Moreover, once the worker is aware of the type of computations being performed, it can generate a random number from a given range (good enough not to raise suspicion of attack) and report it as the output. For implementing the above assumptions, the function $g(x)$ is the simple random number generator function in JAVA. The malicious participant uses this less costly function randomly to introduce error in the output of the main computations.

4 Proposed Scheme

As stated in the Introduction, the proposed scheme is a hybrid of the sampling and checkpointing based schemes. There are three steps in the proposed scheme:

1. Preprocessing the input data
2. Injecting strategic input inside the preprocessed input using appropriate sampling strategy as part of the detection procedure(*Sampling*)
3. Feeding the preprocessed data along with the injected strategic input to the MapReduce cluster and obtaining output (*Checkpointing*).

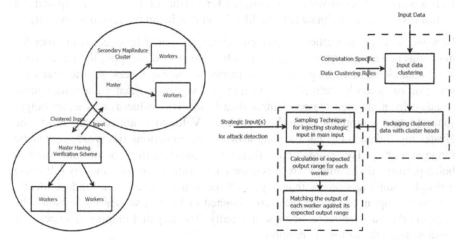

Fig. 1 Overview of the Proposed Scheme

4.1   Preprocessing the Input Data

In most of the detection schemes like [12], preprocessing means performing the same or related set of computations on the input data. The preprocessing in our proposed scheme is the general clustering of input data depending upon the number of mappers in the system. This is done so that data in a given range is given to each mapper and each mapper gets roughly equal share of input for processing. This also helps in sampling and targeting a particular mapper/reducer for verification. Moreover, the clustering of input data is also done by a separate MapReduce cluster thereby matching processing speed between primary and secondary cluster. We refer to it as *Secondary Cluster*.

4.2   Injecting Strategic Input

In this phase, we try to follow an appropriate sampling procedure to select the participant(s) to be targeted. The system needs a small fraction of total number of participants to be verified at a time. Each of this small number of participants must be selected at random with equal likelihood of selecting each participant from the total list of participants. The random process has a property of replicating its output [15]. This can be a problem if there is small number of trials. This can lead many workers not even being verified by the system. So, in addition to not being purely random in nature, the random sampling used for injecting the strategic input should take care of the following two likelihoods:

- If a worker or set of workers is selected for verification in current step and the test is negative (legitimate action), likelihood that it will behave maliciously the next time. *(s)*
- If a worker or set of workers is selected for verification in current step and the test is negative (legitimate action), likelihood that it can be trusted forever. *(t)*

Now we state the properties of strategic input that should be injected in order for malicious behaviour of any of the workers being detected. Suppose there are n total input data elements to be processed, m mappers and r reducers. As the next step, selection rules can be defined for selecting data value(s) d_i that are at a maximum distance from targeted clustered input data head. This value d_i acts as an output space for the computations on the input data. Value(s) v_i are injected inside the cluster input data such that the output of the computations lies inside a specific range around d_i. The value v_i must therefore depend on the type of computations being performed The analysis results shown in section 5.3 are obtained by following a simple sampling strategy that targets all the workers evenly and independently (random sampling). The mappers are assigned tasks in a sequential fashion. The mappers then determine reducers sequentially. The targeted mapper is selected at random using the following relation:

$$P_{global} = P_{mapper} + n/m \times (M - 1) \tag{1}$$

$$P_{global} = P_{reducer} + n/m \times (M - 1) + (R - 1) \times (n/mr) \tag{2}$$

Where,
P_{global}: Position at master node
P_{mapper}: Position at that particular mapper
$P_{reducer}$: Position of the reducer
M: mapper number in which the position is present locally
R: reducer number in which the position is present under a particular mapper(eg: if we have mapper m_1 and m_2. Reducer to m_1 are r_{11}, r_{12} and to m_2 are r_{21}, r_{22}. So if the data position is present in r_{22},then R will be 2 not 4)
n: Total number of input data items at master node
m: Total number of mappers
r: Total number of reducers present under a particular mapper

4.3 Obtaining the Output

The overall output of the computations can be collected by the master and since the strategic input was injected, results will slightly deviate from desired. This phase consists of the steps taken for separating the effect of strategic input from the legitimate output. It may be required to checkpoint various steps during execution. For timely delivery of results, it is required that either the output of strategic input be discarded or the output be recalculated using a separate computing unit.

5 Experimental Evaluation

To evaluate the proposed scheme, Hadoop implementation of MapReduce is used. Experiments carried out in this research aim at analyzing the detection rate of the attacks using the proposed technique.

5.1 Experimental Setup

The experimental setup for this research has two small clouds of 6 machines each. Of these, 5 machines act as the workers and offer MapReduce services and one host acts as a master. Each machine is a intel core i3 processor having 512 MB main memory and 20 GB hard disk. The main Hadoop cluster preprocesses the input as explained in section 4.1 using MapReduce version of K-means algorithm and forwards the preprocessed data to the main cluster to operate upon using shell scripts. So, the proposed scheme operates on default Hadoop configuration. For analysis, we implemented MapReduce version of standard deviation calculation and check-pointed various temporal segments of the processing and collect the data. The computation provider modifies the computations such that the malicious worker "learns "from the previous results of the same computations and next time returns a random value in the range determined by the maximum and minimum output of the previously computed results.

5.2 Threat Analysis

The attacker model capabilities assumed for this paper have already been discussed in section 3. This section presents analysis of threat the attacker model poses to the MapReduce setup. While analysing the threat, we assume that the system does not have any knowledge of the type of input and only the main cluster (not the secondary

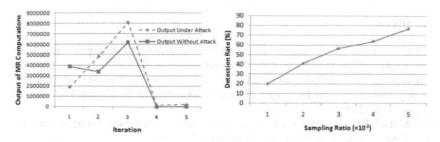

Fig. 2 (a)Analysis of threat due to Attack Model and (b)Variation of Detection Rate of the Scheme with Sampling ratio

cluster) processes the input. The threat is analysed in terms of the units of variation of the obtained output from the desired output.

5.3   Performance Analysis of Verification Mechanism

Performance analysis of the detection scheme is measured in terms of ratio of the number of times the attack was detected and the total number of times the attack happened. Frequency of the attack and frequency of result check are to be appropriately set for better analysis. By far, we assumed that the attacker node and node whose result is to be verified is randomly selected. The result shown in figure 2 is expected to improve even more when the parameters s and t (as explained in section 4.2) are properly modelled. Horizontal axis consists of sampling ratio, which is the number of strategic input data injected per unit the total number of input data.

6   Conclusions and Scope

The problem of result verification has been addressed for wide range of computing platforms. Adapting result verification schemes for statistical computations in MapReduce environment is not straightforward. The scheme proposed in this paper detects the malicious behaviour by a host in case either the computation provider or the participating hosts are untrusted. The scheme employs a secondary cluster performing the computation-independent tasks for preprocessing the input. That means the secondary cluster does not have to trust the computation provider as is the case when the computations are exactly replicated on the secondary cluster. Since the secondary MapReduce matches the speed of processing by the main cluster, the timeliness of result delivery is not affected. The proposed scheme was evaluated on Hadoop implementation of MapReduce. Analysis of threat due to the attacker model assumed in this paper and detection rate achieved by the proposed scheme are also presented. However, the scheme needs refinement so that it works for more complex MapReduce computations. Making the sampling strategy and methodology of injecting strategic input into the main input more powerful remain as the ongoing future work.

References

1. Dean, J., Ghemawat, S.: MapReduce: simplified data processing on large clusters. Communications of the ACM 51(1), 107–113 (2008)
2. Zhao, S., Lo, V., Gauthier Dickey, C.G.: Result verification and trust-based scheduling in peer-to-peer grids. In: Fifth IEEE International Conference on, August 31-September 2, pp. 31–38 (2005)
3. Sarmenta, L.F.G.: Sabotage-tolerance mechanisms for volunteer computing systems. Future Generation Computer Systems 18(4), 561–572 (2002)
4. Domingues, P., Sousa, B., Silva, L.M.: Sabotage-tolerance and trust management in desktop grid computing. Future Generation Computer Systems 23(7), 904–912 (2007)

5. Golle, P., Stubblebine, S.G.: Secure distributed computing in a commercial environment. In: Syverson, P.F. (ed.) FC 2001. LNCS, vol. 2339, p. 279. Springer, Heidelberg (2002)
6. Golle, P., Mironov, I.: Uncheatable distributed computations. In: Naccache, D. (ed.) CT-RSA 2001. LNCS, vol. 2020, pp. 425–440. Springer, Heidelberg (2001)
7. Indrajit, R., Srinath, T.S., Kilzer, A., Shmatikov, V., Witchel, E.: Airavat: Security and Privacy for MapReduce. In: NSDI, vol. 10, pp. 297–312 (2010)
8. Sander, T., Tschudin, C.F.: Protecting mobile agents against malicious hosts. In: Vigna, G. (ed.) Mobile Agents and Security. LNCS, vol. 1419, pp. 44–60. Springer, Heidelberg (1998)
9. Wei, W., Du, J., Yu, T., Gu, X.: SecureMR: A Service Integrity Assurance Framework for MapReduce. In: Annual Computer Security Applications Conference, ACSAC 2009, December 7-11, pp. 73–82 (2009)
10. Zhao, S., Lo, V., Gauthier Dickey, C.: Result verification and trust-based scheduling in peer-to-peer grids. In: Fifth IEEE International Conference on Peer-to-Peer Computing, P2P 2005. IEEE (2005)
11. Szajda, D., Lawson, B., Owen, J.: Hardening functions for large scale distributed computations. In: Proceedings of the 2003 Symposium on Security and Privacy. IEEE (2003)
12. Xiao, Z., Xiao, Y.: Accountable MapReduce in cloud computing. In: SCNC 2011 (2011)
13. Hadoop Tutorial,
   `http://public.yahoo.com/gogate/hadoop-tutorial/`
   `start-tutorial.html`
14. Huang, C., Zhu, S., Wu, D.: Towards Trusted Services: Result Verification Schemes for MapReduce. In: 2012 12th IEEE/ACM International Symposium on Cluster, Cloud and Grid Computing (CCGrid), May 13-16, pp. 41–48 (2012)
15. Kroese, D.P., Taimre, T., Botev, Z.I.: Uniform Random Number Generation. In: Handbook of Monte Carlo Methods, ch. 1, p. 772. John Wiley & Sons, New York (2011)

Quantifying Direct Trust for Private Information Sharing in an Online Social Network

Agrima Srivastava, K.P. Krishnakumar, and G. Geethakumari

Abstract. Online Social Networks (OSNs) are actively being used by a large fraction of people. People extensively share a wealth of their information online. This content if retrieved, stored, processed and spread beyond scope without user's consent may result in a privacy breach. Adopting a coarse grained privacy mechanism such as sharing information to a group of "close friends" or the strong ties of the network is one of the solutions to minimize the risk of unwanted disclosure but this does not fully contribute in the process of protecting privacy. There is a high probability for an unwanted information disclosure even if the information is shared only with the strong ties. In most of the privacy literature while building trust the online sharing behavior is never looked into consideration. Hence, in this paper we propose and implement a privacy preserving model where such unwanted and unintentional private information disclosures could be minimized by further refining the trusted community of strong ties with respect to their privacy quotient.

Keywords: privacy, direct trust, strong ties, online social networks.

1   Introduction

The fast adoption of OSN is accompanied with problems like disclosure and unwanted access of sensitive data which results in an individual's privacy breach [1]. Privacy and trust are highly related to each other. Trust is defined as a measure of confidence that an entity or entities will behave in an expected way despite the lack of ability to monitor or control the environment in which it operates. It plays a role in the formation of social communities

Agrima Srivastava · K.P. Krishnakumar · G. Geethakumari
Department of Computer Science and Information Systems
BITS Pilani, Hyderabad Campus, Hyderabad, India

© Springer International Publishing Switzerland 2015   21
R. Buyya and S.M. Thampi (eds.), *Intelligent Distributed Computing*,
Advances in Intelligent Systems and Computing 321, DOI: 10.1007/978-3-319-11227-5_3

as well helps determine how information flows in the network [2]. In order to prevent unwanted disclosures it is important to form a trusted community where all the members can share their thoughts and opinions without the fear of privacy breach.

Problem Statement: To form a trusted community we need to quantify and measure the trust for each individual. The amount of trust user X has on user Y can easily be collected by monitoring X's experience with Y, measuring X's attitude information for Y or by tracking X's behavior for Y over time [3]. Interactions are a measure of trust [4]. The users often interact with their strong ties and share most of their information with them. Though a community of people with strong ties is a highly trusted one, but when privacy is a concern sharing of information with these strong ties without considering their online sharing behavior can result in an unwanted and unintentional information disclosure. The problem here is to minimize these insider privacy attacks where most of the private information is spread because of the unconcerned users within the trusted community.

Rest of our paper is organized as follows; Part 2 describes the related work. In Part 3 we first discuss the overview of the proposed privacy preserving model and then give a detailed explanation for the same. In Part 4 we run simulations and discuss the results obtained. Finally in Part 5 we conclude and give an outline for future work.

2 Related Work

Liu et al [5] have provided an intuitively and mathematically sound methodology for computing the privacy scores of users in the OSNs by making use of the Item Response Theory model. They have used the two parameter logistic model to calculate the privacy scores of individuals. Kumarguru et al [6] have described the methodology, questions and the results obtained used by Westin to create a privacy index. Lind et al [7] have proposed measures like spreading factor and spreading time which are accessible neighborhood around the node and the minimum time to reach such neighborhood respectively. These properties give an insight about the private information spreading in the network. Shaw et al [8] have revealed that how the information that they had passed along one edge can affect the strength of the other edges too. They conducted experiments and concluded that gossip decreases the network clustering and the average node degree.

Buskens et al [9] discusses the explanations for the emergence of trust where the nodes are regularly informed about the other node's behavior and label them as untrustworthy. Adali et al [4] evaluate trust on the basis of communication and interaction amongst the members in the social network. They calculate the behavioral trust as conversation and propagation trust.

There are different hybrid trust models that uses the interactions and social network structure for computing trust. Trifunovic et al [10] have proposed approaches like the explicit and the implicit social trust where explicit social trust is derived from the strong ties based on the frequency of interactions and the implicit social trust is derived on the basis of frequency and the duration of contact between the users. The trust calculated is directly proportional to the weight of the relationship. In our work we also look into the online sharing behavior of the nodes, compute the privacy quotient (PQ) for them and refine the trusted community to reduce the unwanted disclosures significantly.

3   The Detailed Explanation of the Proposed Model

In figure 1 we show a trust aware model that calculates trusts on the basis of the weight of the relationship.

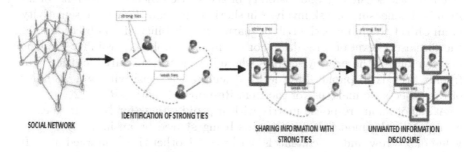

Fig. 1 Model that does not consider the online sharing behavior of strong ties

In Figure 2 the proposed privacy preserving model is shown where before finalizing the strong ties their online sharing behavior is also taken into consideration. Using this the strong ties are classified as unconcerned, pragmatics

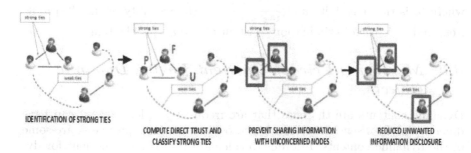

Fig. 2 The proposed privacy preserving model that considers the online sharing behavior of strong ties

and fundamentalist. We depict them as U, P and F respectively. After classifying the unconcerned strong ties are filtered and the information is shared
with the refined trusted community of strong ties. Using this approach the
unwanted disclosures are reduced up to a great extent. In the following sub-
sections we prove the same.

3.1   Measuring Privacy Quotient for the Static Objects

In an OSN the static profile items are the ones that are not changed too frequently and have a predefined or limited set of values. Some of the examples of
static profile items are hometown, current-town, birthday, relationship status,
contact number, job details, email etc. Static Privacy Quotient i.e. PQ(Static)
is a measure of the privacy of an individual [11]. Privacy is abstract and in
order to compute the static privacy quotient for static objects, we need to
calculate the sensitivity and visibility of the static objects. Sensitivity of an
item is the measure of risk involved in sharing an item. Higher the sensitivity
of an object the more is the risk in sharing it. Visibility of an object defines
the disclosure level of an object. For eg. in Facebook the visibility levels is
synonymous to the privacy settings where it ranges from only me, friends of
friends, friends to public. In order to calculate the sensitivity and visibility
of the objects we make use of the Item Response Theory (IRT) [12].

A dichotomous response matrix with n profile items for N users goes as
an input to the model. If an item i is being shared by an individual j the
value of jth row and i^{th} column is marked as 1 otherwise is marked as 0. If
the ability of each of the individual is θ_j then, $P_{ij}(\theta_j)$ is the probability that
the jth individual with an ability θ_j will share an item having an index i [12].
PQ(Static) is the product of *sensitivity* and the *visibility* of the profile items
[5]. Privacy quotient of a user can be calculated as

$$PQ(Static)_j = \sum_i \beta_i * \frac{1}{1 + e^{\alpha_i(\theta_j - \beta_i)}} \qquad (1)$$

where β_i is the sensitivity and $\frac{1}{1+e^{\alpha_i(\theta_j-\beta_i)}}$ is the visibility of the ith profile
item and α_i is the discrimination constant of the ith profile item.

3.2   Measuring Privacy Quotient for the Dynamic Objects

Dynamic contents are the ones that are frequently uploaded and can have
different levels of sensitivity. User's photos, videos, links, posts etc. are some
of the dynamic contents. In order to calculate the privacy quotient for dynamic objects we need to look at the ratio of objects kept as private with the
ratio of objects shared to public.

$$PQ(dynamic) = \frac{CPr}{CP} \tag{2}$$

Here CPr is the count of objects kept as private and CP is the count of objects shared to Public. The users having this ratio less than 1 are the ones who share a high percentage of objects to public in comparison with what they keep as private. We can either use PQ(Static)or PQ(dynamic) based on the privacy requirement of the target user.

3.3 Forming a Trusted Community of Strong Ties Based on the Interactions

The nodes in an OSN exchange information with the other nodes to build and maintain the social capital. The combination of amount of time, the emotional intensity, the intimacy and the reciprocal service characterizes a tie. There are *strong ties* and *weak ties* in the network. The connection between the individuals who belong to the distant areas of the social graph are known as the weak ties. In contrast to this the strong ties are the trusted and known individuals. Any OSN comprises of sets of nodes and edges where the nodes are the actors or individuals and the edges signify the relationship between them. The nodes interact and communicate with each other. These interactions are a good indicator of social relationships and hence a good indicator of conversational trust [4]. The strong ties of any node X are the nodes in the network with whom the node X interacts frequently. These nodes usually form a community of close friends of the user.

3.4 Quantifying the Direct Trust of All the Strong Ties in the Network

A lot of OSNs use the concept of "friends" or "close-friends" for privacy control. Though this reduces the risk of disclosure but it is extremely coarse. We require a fine-grained privacy control management using which not only the users would have the privilege to share objects to their close friends but also to their privacy aware trusted friends. This can help the users to regulate the data that is shared online without limiting the value of social capital. In order to achieve this we need to compute the direct trust between the two nodes having an edge [13]. Direct trust is the trust value obtained with the direct relationship. If s and t are the two nodes then the direct trust between s and t can be calculated as

$$DT(s,t) = \frac{W(R_i)}{N_s} * pfactor(s,t) \tag{3}$$

Here $W(R_i)$ is the weight of the relationship between s and t. A strong tie interacts more with the target user in comparison with a weak one and hence the strong tie will have a higher value of $W(R_i)$ than the weak tie.

N_s are the total number of edges a node s has.

pfactor of a node s for node t can be defined as

$$pfactor(s,t) = PQ(t) * (1 - PQ(s)) \tag{4}$$

3.4.1 Derivation of pfactor

The probability that node t will hide the information and node s shares it to t is the pfactor(s,t) which is given as:

$$pfactor(s,t) = Prob(t\ hides) \bigcap Prob(s\ shares) \tag{5}$$

$$Prob(t hides) \bigcap Prob(s shares) = (Prob(t hides)|P(s shares)) * Prob(s\ shares) \tag{6}$$

Event 1 : Node s sharing the information to t and *Event 2* : Node t hiding the information.

Event 1 and Event 2 are independent events. So the final equation can be rewritten as

$$Prob(t\ hides) \bigcap Prob(s\ shares) = Prob(t\ hides) * P(s\ shares) \tag{7}$$

Privacy quotient (PQ) is nothing but the probability that determines the intensity of information hiding. We can replace the probabilities with the respective privacy quotients of the nodes.

Probability of t hiding information is the PQ of t and the probability of s sharing an information is (1 - PQ(s))

$$Prob(t\ hides) \bigcap Prob(s\ shares) = PQ(t) * (1 - PQ(s)) \tag{8}$$

Adding the pfactor while computing the direct trust will also consider the online sharing behavior of the receiving node t before s builds a trust value for it.

3.5 Classifying Nodes on the Basis of Privacy and Filtering Out the Unconcerned Users from the Trusted Community

Dr. Alan Westin has conducted over 30 privacy surveys and created one or more privacy indexes to summarize his results and show the trends in privacy concerns. Westin has classified public into three categories

- High and the Fundamentalist : These are the people who are cautious and worried about their privacy. They are distrustful of the organizations and choose privacy over any form of consumer service benefits.
- Medium and the Pragmatics : These are the ones who believe that business organizations should "earn" the trust rather than assuming that they have it. They look at the benefits provided to them in comparison with the degree of intrusiveness of their personal information.
- Low and the Unconcerned : These people trusts the organizations with the collection of their private information and are ready to use the customer service benefits in exchange of their personal information.

Using the direct trust (DT) calculated from equation 3 we categorize all the strong ties as unconcerned, pragmatics and fundamentalist. We use k means clustering to determine the two thresholds.

Table 1 Classification of strong ties according to Westin's classification

DT $\geq$ 0 and DT < threshold 1	Unconcerned Nodes
DT $\geq$ threshold 1 and DT < threshold 2	Pragmatics Nodes
DT $\geq$ threshold 2	Fundamentalist Nodes

After identifying the unconcerned nodes within the trusted community of friends, we refine the community by filtering out them. In the next section we validate the algorithm on real Facebook datsets and prove the same.

4   Results and Discussion

4.1   Simulation Set Up

We obtained the data sets from Stanford Network Analysis Project [14] for Facebook which is a popular OSN. Table 2 lists the complete statistics of the

Table 2 Data Statistics of the collected data

Nodes	4039%
Edges	88234%
Nodes in largest WCC	4039 (1.000)%
Edges in largest WCC	88234 (1.000)%
Nodes in largest SCC	4039 (1.000)%
Edges in largest SCC	88234 (1.000%
Average clustering coefficient	0.6055%
Number of triangles	1612010%
Fraction of closed triangles	0.2647%
Diameter (longest shortest path)	8%
90-percentile effective diameter	4.7%

Table 3 Data statistics of the collected data

	Number of Nodes	Average Degree Network Diameter	Average Clustering Coefficient
Network 1	1034	51.739	.534
Network 2	224	28.5	.544
Network 3	150	22.573	.67
Network 4	168	19.764	.534
Network 5	786	35.684	.476
Network 6	747	80.388	.635
Network 7	534	18.026	.544
Network 8	52	5.615	.462
Network 9	333	15.129	.444

different networks that were studied. Table 3 lists some of the properties like the number of nodes, clustering coefficient, average node degree etc. for each network. We performed two sets of simulations with the aim of measuring the private information spread.

- *Simulation 1:* Measure the private information spread through the uncon-cerned, pragmatics and fundamentalist strong ties of the target user.
- *Simulation 2:* Measure the private information spread through the prag-matics and fundamentalist strong ties of the target user.

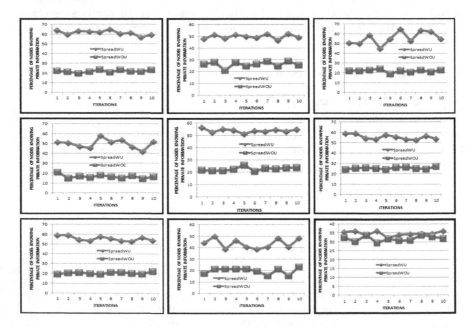

Fig. 3 Graph showing a comparison of private information spread with and with-out including the unconcerned strong ties

In Figure 3 the blue (rhombus) line indicates the spread of private information including the unconcerned nodes and the red line (square) indicates the spread of private information without including the unconcerned nodes. We measured the private information spread with list of networks mentioned in table 3. Each network was studied for 10 iterations. The graphs show the percentage of nodes knowing the private information with every iteration. The Independent Cascade Model [15] was used to model this information spread from the source to the other nodes in the network.

Observation: It was clearly observed that while sharing the private information with the trusted strong ties if the unconcerned users were excluded the private information aware nodes were reduced up to a greater extent.

5 Conclusion and Future Work

Trust is an important element in OSN. Offline people do not share their information to anyone and everyone. They selectively disclose their private information. The same behavior is difficult to replicate online. Now-a-days some of the social networking sites allow the users to share their sensitive information only to their close and trusted friends. The strong ties in the network are the users who interact often with the target user. These strong ties are the close and the trusted friends of the target user. Where privacy is a concern, friends who are close to us may not have a good privacy quotient. Hence in this paper we propose and implement a privacy preserving model that calculates the direct trust between two nodes after considering its online sharing behavior. We then identify and filter out the unconcerned strong ties within the network and reduce the unwanted spread caused by the members of the trusted community. In future we will be exploring more on the area of trust and will try to quantify the indirect trust and use it to preserve privacy.

References

1. Gross, R., Acquisti, A.: Information revelation and privacy in online social networks. In: Proceedings of the 2005 ACM Workshop on Privacy in the Electronic Society, pp. 71–80. ACM (2005)
2. Singh, S., Bawa, S.: A privacy, trust and policy based authorization framework for services in distributed environments. International Journal of Computer Science 2(2) (2007)
3. Sherchan, W., Nepal, S., Paris, C.: A survey of trust in social networks. ACM Computing Surveys (CSUR) 45(4), 47 (2013)
4. Adali, S., Escriva, R., Goldberg, M.K., Hayvanovych, M., Magdon-Ismail, M., Szymanski, B.K., Wallace, W.A., Williams, G.: Measuring behavioral trust in social networks. In: 2010 IEEE International Conference on Intelligence and Security Informatics (ISI), pp. 150–152. IEEE (2010)

5. Liu, K., Terzi, E.: A framework for computing the privacy scores of users in on-line social networks. In: Ninth IEEE International Conference on Data Mining, ICDM 2009, pp. 288–297. IEEE (2009)
6. Kumaraguru, P., Cranor, L.F.: Privacy indexes: A survey of westin's studies (2005)
7. Lind, P.G., da Silva, L.R., Andrade Jr., J.S., Herrmann, H.J.: Spreading gossip in social networks. Physical Review E 76(3), 036117 (2007)
8. Shaw, A.K., Tsvetkova, M., Daneshvar, R.: The effect of gossip on social networks. Complexity 16(4), 39–47 (2011)
9. Buskens, V.: Social networks and trust, vol. 30. Springer (2002)
10. Trifunovic, S., Legendre, F., Anastasiades, C.: Social trust in opportunistic networks. In: INFOCOM IEEE Conference on Computer Communications Workshops, pp. 1–6. IEEE (2010)
11. Srivastava, A., Geethakumari, G.: Measuring privacy leaks in online social networks. In: 2013 International Conference on Advances in Computing, Communications and Informatics (ICACCI), pp. 2095–2100. IEEE (2013)
12. Baker, F., Kim, S.-H.: Item response theory: Parameter estimation techniques, vol. 176. CRC Press (2004)
13. Li, N., Najafian Razavi, M., Gillet, D.: Trust-aware privacy control for social media. In: CHI 2011 Extended Abstracts on Human Factors in Computing Systems, pp. 1597–1602. ACM (2011)
14. McAuley, J.J., Leskovec, J.: Learning to discover social circles in ego networks. In: NIPS, vol. 272, pp. 548–556 (2012)
15. Kempe, D., Kleinberg, J., Tardos, É.: Maximizing the spread of influence through a social network. In: Proceedings of the ninth ACM SIGKDD International Conference on Knowledge Discovery and Data Mining, pp. 137–146. ACM (2003)

A Heuristic for Link Prediction in Online Social Network

Ajeet Pal Singh Panwar and Rajdeep Niyogi

Abstract. As we know due to advancement in technology it is very easy to be in connection with others. People interact with each other and they create, share and exchange information and ideas. Social network is one of the most attracted areas in recent years. Link prediction is a key research area. In our proposed method we study link prediction using heuristic approach. Most of the previous papers only considered the network topology, they didn't consider the nodes properties individually, and they treated them only as passive entity in graph and using only network properties. But in our proposed method we will consider different parameter of nodes that define the behavior of nodes and one important issue that we will consider in our method is "**New researchers**" because they are willing to get help in identifying potential collaborators. Thus our focus will also on "**New researchers**" and we would also like to have a quantitative analysis of the performance of the different existing methods and to study some domain specific heuristics that would improve the degree of prediction.

Keywords: Link prediction, Co-authorship Networks, Co author prediction, Social Network.

1   Introduction

Social network can be defined as a network which is made up of individuals or organizations, which are interconnected, Here individuals or organizations represents nodes in network and link between them represents relationship or connection . To represent social groups we can use social network which shows how the group influence by others entities present in group and how its behaviors changes over time.

Ajeet Pal Singh Panwar · Rajdeep Niyogi
Department of Computer Science & Engineering, IIT Roorkee, Roorkee-247667
e-mail: {ajeetapsp8,rajdeepniyogi}@gmail.com

© Springer International Publishing Switzerland 2015   31
R. Buyya and S.M. Thampi (eds.), *Intelligent Distributed Computing*,
Advances in Intelligent Systems and Computing 321, DOI: 10.1007/978-3-319-11227-5_4

Here an important question arises how to predict a link in a social network which we can say is a relationship or connectivity between nodes. This is called **Link prediction** problem, we propose an algorithm to predict the links which are likely to appear in network in future. There can be many different kinds of social network but we are considering the network of scientist or authors, if two authors have published a paper together then their interaction is denoted by an edge or a link in network. Using our algorithm we will predict the new interaction between the nodes of social network those don't have collaboration before by analyzing the snapshot of network. Prediction of new link is not a easy task but by analyzing the given properties of network and behavior of existing nodes, new link can be predicted which are likely to appear in future.

2   Related Work

In social network analysis we can analyze the network by considering in degree, out degree, shortest path between nodes etc but this information is not sufficient because we know that graph changes dynamically so we have to analyze the internal structure of network, and for this we have to consider some specific properties of network such as common neighbor , preferential attachment etc with some semantic information such as research interest, year of publication and many others. So many authors have contributed in the field of social network. They have given so many algorithms for predicting the future links in social network. Researchers shows the intriguing similarity in structural properties of network of different types by hypothesis that a node which is highly connected increases its connectivity faster in compare to a node which is less connected. It means the rate at which nodes makes connection is directly proportional to node's degree, such phenomenon is called as preferential attachment[2].Barab asi Alber(BA) model shows that probability distribution weighted by node degree can be used to predict new links.Some researchers have shown the extension of BA models ,for example display a high degree of transitivity , it means there is high possibility that two nodes will connect to each other if they have some common acquaintances. This phenomenon is known as clustering . Watts and Strogatz measured the clustering in many types of real world networks, for this we have to calculate the clustering coefficients "C".

C= 3 * (Number of triangle on the graph)/(Number of connected triples)

Here triangle represent set of three closed triplets. A triplet can be defined as set of three nodes which are connected by either 3 closed triplet or 2 open triplet.Some class of models which grows power-law network works on some local rules. for example random walk[3] and common neighbor[2] , where node select new nodes by walking a random walk or by considering the common friends.Some class of models takes the snapshot of network at different point of time and then makes comparison for examine the network growth process. Pallavi[10] has used heuristic approach in which she used Knowledge dependency graph and some specific properties of nodes to predict links. Newman[2] found the evidence of preferential attachment when examined the two collaboration network.Jeon et al[4] found that nodes makes link in proportion of their degree.Lada A.Adamic , Eytan Adar [5] shows for link prediction similarity between nodes can be measured by analyzing text, link, mailing list etc. It is well known the more items are common in two people the more they are likely to be friend, but some time items that are unique to few number of users are more weighted than commonly occur items.[11] presents a methodology for specifically examining geographical effects on intra-national scientific collaboration and it shows that when the distance between the collaborative partners increases then the collaboration between them decreases exponentially.[8] shows that semantic description can be used to find collaboration between authors. If we can find to what extent an author is expertise in specific field, one can effectively used this knowledge to find authors with compatible expertise.[7] method used some semantic attributes of the nodes like title and abstract information, it also uses node properties like local density and community affiliation information more intensely.

3   Proposed Approach

As we know social network changes and grow rapidly with the addition of new links in network. This is because of interaction in the underlying network, how these network evolves is really a challenge .Link prediction is analysis of social network evolution, we have given a snapshot of a social network at a particular moment of time 'X' and we have to predict the new link in social network after some moment of time 'Y'. In our proposed method we are using the network of scientists for representing the graph. In graph each node represents a specific author and edge between two nodes represents the collaboration between two nodes (authors). Our aim is to discover new edges in graph by using the earlier collaboration between nodes (authors) and some specific properties of nodes (authors) for example their research area, current publication, co authors etc. we have collected our data set by visiting different authors websites, their research papers and then made a network of coauthors. In graph we are considering only those paper which are published only between time period 2003 to 2014 . Given time period is divided into two time periods one is training and another is testing. Using training period (2003 to 2010) we create a graph where nodes are connected if they have published paper together in training time period ,Testing period(2011 to 2014) is used for checking whether the edges predicted by the analysis of data in training period is correct or not.

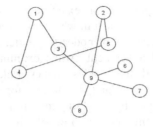

Fig. 1 Graph 'A' created at time T1

Let assume we have given graph 'A' at time T1, which is shown figure 1, After addition of new edges in graph 'A' it looks like graph 'B' at time T2 , which is shown below in figure 2.

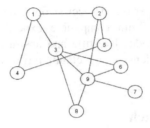

Fig. 2 Graph 'B' created at time T2 (T2>T1)

There are many edges in graph 'B' which were not present in graph 'A'. The graph 'A' and 'B' represent graphs without any prediction according to our collected data but the graph 'C' in figure 3 is created by using link prediction algorithm. In graph 'C' it may possible that prediction of edges is greater than the actual number of edges in graph 'B' and the ration of both can be used as performance measure. For example in graph 'B' the total number of new edges in graph is 'a' and total number of edges in graph 'C' is 'b' but out of 'b' only 'c' prediction are correct, so the performance percentage will be calculated as follows:

Performance= ((Total number of correct prediction)/ (Total number of prediction))*100 = (c/b)*100;

These performance measures can be used to compare the different methods for link prediction in social network. There are so many existing methods which are based on the network topology. In our proposed method we show that not only network topology but nodes information can also be used as effective tool for link prediction, because here node is an author who has some specific properties. We are considering the co authorship domain for illustration of our proposed method.

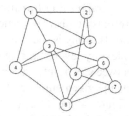

Fig. 3 Graph 'C' predicted after analysis of data at T1 time

In our method initially we have given a graph G= (V, E) which represent the social network of authors , the graph is undirected because the relation is symmetric.Here we want to create a graph G`=(V,E`) where E` is the number of edges which is union of old and new edges EUE_new , E_new represent the set of new edges.

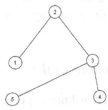

Fig. 4 Graph before transitive closure

In our method we use the concept of transitive closure, which gives some edges that may occur in future.

Transitive Closure: Transitive closure of a binary relation R on a set X is the transitive relation R+ on set X such that R+ contain R and R+ is minimal. A binary relation is transitive whenever an element 'x' is related to element 'y' and element 'y' is related to 'z' then 'x' is also related to 'z'. Actually the concept of transitive closure suites in our proposed method, if pair of authors 'x' and 'y' have published paper together and author 'y' has also published paper with author 'z' then it is possible that both 'x' and 'y' will come closer and publish paper.

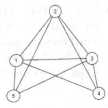

Fig. 5 After transitive closure

Maximum number of edges in graph using transitive closure cannot more than $((V*(V-1)/2)-E)$. Here 'V' represents number of vertices and E represents number of edges initially given in graph. But this is the maximum number of connection possible in graph, it may possible only few of them will occur in future, this is not always true that connection between two authors can be obtain using transitive property so it necessary to refine the given number of edges.

The number of edges in figure 4 are (1,2),(2,3),(3,4),(3,5) after transitive closure we have ((1,2),(1,3),(1,4),(2,5),(2,4),(2,3),(3,4),(3,5)) number of edges in figure 5 out of which some of edges will occur in future. For filter the number of edges we can use the properties of network structure and some attributes of nodes in network. Here we will define some parameters that can help us to predict the link prediction between authors. These parameter may include the network structure but as we know our graph is not simply a graph ,it is a representation of social network it means each node is an author and each node behaves in different manner for example each node may have different research interest , different co authors, worked with different authors etc. In our proposed method we have categorize our approach in two different ways whether there is a known person between two authors or not.

3.1   Parameters for Predicting Links

a) If authors do not have common known person, then we have consider the following parameter

1: Research Interest: Each author has different research interest. But we confine our interests in computer science subjects. For example node '1' (author) interest are "**machine learning**" and "**robotics**" and he has published paper recently in 2010. Lets another node '5' interest are "**computer vision**" and "**robotics**" ,here we can see they have "robotics " as a common interest .But having common research interest is not sufficient so we have to find whether two authors are currently active or not. For this we check their recent publication. If they have published paper recently and year gap of paper publication is very less then we can consider them as probable edge in graph. For example node '1' has published in 2010 and node '2' has published in 2008 ,here their year gap is only 2 year and very close to 2010(node '1' publication is 2 year old and node '2' publication is latest)

2: Collaboration between Co Authors: It is always good if authors have similar interest and are currently active but we also know that their co authors will also have same interest. It may possible that their co authors collaborate themselves. If their collaboration exceeds a particular threshold value, then we can consider their collaboration as probable edge in graph, for example we will assume that node '1' and node '2' will also collaborate.

3: Institution: Authors having common interest and currently active are always important for collaboration but if these authors also belong to same institution or organization then their chances of collaboration increases dramatically. For example node '1' and node '2' belongs to same university "**university of oxford**" having same interest and currently active can be considered as probable edge in graph.

4: Project: Sometime authors had done work together but they did not collaborate for publication of papers so these pairs of authors also have good chances for collaboration, for example suppose in an institution a student is under guidance of his supervisor , he has completed his project under his supervisor .As it is known that both of these have worked together, they may have common interest and they also well aware of skills and area of expertise of each other so in future they may prefer to collaborate together.

5: Conferences: Generally authors publish and presents their paper in conferences, it can be used as effective way to know the area of interest of others authors who are attending the specific conference, for example if node '2' and node '3' are attending the conference on *"Advances in Neural Information Processing Systems"* and if any one of them have presented his paper in this conference then other one can easily find whether the presenting person is of his interest or not and this can be used as probable edge in graph.

6: Paper Citation: Paper citation can be used to find collaboration between authors. Usually author takes the references of other author's papers, it means they shows that their work is using some concept from referenced papers for example node '1' is referencing the papers of node '5' continuously from last few year, node '5' can use this information to decide whether node '1' is of his interest or not and this can be used as a probable edge in graph.

7: Knowledge Dependency Graph: Some time research interest of authors do not match, for example plan synthesis and program analysis do not have common name but we know that plan synthesis is related to program synthesis thus also related to program analysis. In order to find such links we define a method known as "Knowledge dependency graph". Let in KDG we have 'K' key words. It is a directed graph KDG=(V,E). For each vertex 'V' there is unique keyword 'K'. Let 'V1','V2' are two vertex in KDG and if there is an edge V1->V2, it represents knowledge in the area represented by 'K1' allow the author to work in the area represented by 'K2'. We can see that this relation is not Symmetric. So if either of **V1->V2** or **V2->V1** edge is in KDG , we consider this pair of authors as probable edge in graph.

b) If authors have common known person, then we have consider the following parameter

1: Recommendation: If authors have common know person then the possibility of their collaboration increases dramatically. Suppose node '1' and node '5' have common node '2', here node '2' and node '1' have done some work before or we can say have collaborated before then node '1' and node '5' may collaborated if node '2' recommend node '5' to node '1'. The possibility of collaboration increases in some special cases for example node '1' and node '2' represents **supervisor** and **student** respectively.

2: High Rank: It is well know that if you want to make new friends in social network you always try to make a friend who has already so many friends and have good image in society for example 'M' is that famous person. It helps us to make new friends because easy access to all friends of 'M'. Same concept we can apply in our problem, if author have a connection with high rank (**high degree**) authors in graph then there is a possibilty that he will collaborate with co authors of high rank authors.

For example if node '4' interest is "**machine learning**" and node '5' interest is "**Artificial intelligence**" there is no keyword matching ,but now suppose node '4' and node '5' has connection with high rank author whose area of interest is "**machine learning**" then there is possibility that node '4' and node '5' will collaborate . In this example we come to know that research interest of node '4' and node '5' may be subset of each other. Remaining parameters like Research, Conference, Institution, Paper citation ,KDG are also applicable for predicting link in case of common known person between authors. Individually these parameters are not sufficient to determine whether two authors will collaborate or not but if we consider all these parameters collectively can be used to predict links in graph. We have to treat these parameters as individual points for collaboration between two authors, chances of their collaboration increases as the sum of these points crosses the particular threshold value for example if two authors have common research interest we can say they may collaborate without knowing about their location or place to which they belongs but if we know that these two authors belongs to same institution and currently active then chances of their collaboration further increases so we can say by considering each parameter prediction of links in graph can be made more accurately compare to considering one parameter at a time for prediction of link. For our method we have to collect necessary information about each individual author which is described as follows.

Adjacency matrix **A[v][v]**, where each Row represents the "Authors" and column represents their "Co authors". A[v][v] shows the relationship between the authors during **training time** period. Rank matrix **Rank[v]**, it shows the degree of each node "Authors " in graph. **Connection[v][v]** matrix, it shows the connection of individual authors with high rank authors. **History[v][v]** , it shows the research interest and recent publication of each author.**TC[v][v]** matrix, this matrix contain the transitive closure of Adjacency matrix **A[v][v]**.

3.2   Pseudo Code for Proposed Method

Link Prediction Using Score Generation

Input: Adjacency matrix say **A[v][v]**, where Row represent the "Authors" and Column represents their "Co authors". **History[v][v]** matrix shows interest and recent publication of authors. **Rank[v]** & **Connection[v][v]** matrix shows individual author connection with high rank authors.**TC[v][v]** shows transitive closure and information about different parameters of each author is also given.

1. Find the transitive closure, it will gives those nodes which were not connected   earlier. it will store results in **TC[][]** matrix.

2. Filter the edges predicted in **TC[][]** matrix: for each edge of pair(i,j)

If (Authors in pair (i,j) do not have any common known author between them)

> **a.** If authors pair(i,j) have done work earlier

>> Increment count(i,j)   // increment count for pair (i,j)

> **b.** If author belong to same institution

>> Increment count(i,j)   // increment count for pair (i,j)

> **End If**

Else (If authors in pair (i,j) do have some common known author)

> **a.** If pair(i,j) of authors have common author who has worked with any one of them. On recommendation of common author one author may collaborate with other author.

>> Increment count(i,j)   // increment count for pair (i,j)

> **b.** If author has connection with high rank authors in social network then he may collaborate with those authors who have same common high rank authors.

>> Increment count(i,j)   // increment count for pair (i,j)

End Else

3. Some parameters are common for both above cases such as common conferences, common interest, currently active, Paper citation, co author collaboration etc. On successful evaluation of each parameter a variable **"count"** increments.

4. If pair of authors do not have common research topics then we should use knowledge dependency graph "**KDG**", if edge is there value of count variable increment. If the value of "count" variable crosses a particular threshold value then we consider this pair as probable edge in graph otherwise not.

3.3   *Experimental Results*

In our proposed method we cannot predict accurately that which pair of authors will collaborate in future. The percentage of prediction is very low because final decision of collaboration depends on author's behavior as he is social person and he may affect by his surrounding which may nullify the effect of all parameters that we have discussed. In our method, we predicted about 186 new collaborations or links in graph but out of these only 16 predictions are correct.

Table showing comparison between different algorithms

Algorithm	# of predictions	# of correct predictions	Performance
Common neighbors method	146	11	7.5 %
Preferential Attachment method	310	13	4.19 %
Heuristic method	197	6	3.04%
Proposed method	186	16	8.6 %

4   Conclusion

The approach use in previous methods are basically based on the network topology of graphs for example common neighbor, node degree etc but these methods are not efficient when graph is sparse .In our method we have taken small graph in which sometime network properties are not sufficient to predict the edges efficiently, therefore we have to consider the analysis of individual nodes which are actually "authors" and behaves actively . There are some new ways to improve our approach that will use structure of graph and also efficiently uses the content based feature, semantic information such as abstract, title, event information to improve the link prediction. Using semantic information we can predict the links in graph even if the graph is sparse which makes it as a useful tool for link prediction.

References

1. Liben Nowell, D., Kleinberg, J.: The link prediction problem for social networks. In: Proceedings of the Twelfth International Conference on Information and Knowledge Mangement, CIKM 2003 (2003)
2. Newman, M.E.J.: Clustering and preferential attachment in growing networks. Physics Review E (2001)
3. Saramaki, J., Kaski, K.: Scale-free networks generated by random walkers. Physica A 341 (2004)
4. Jeong, H., Neda, Z., Barabasi, A.-L.: Measuring preferential attachment for evolving networks. Europhysics Letters (2003)
5. Adamic, L.A., Adar, E.: Friends and neighbors on the web (2003)
6. Fire, M., Tenenboim, L.: Link prediction in social networks using computationally efficient Topological feature. In: 2011 IEEE International Conference on Privacy, Security Risk (2011)
7. Sachan, M., Ichise, R.: Using Semantic information to improve Link prediction results in network datasets. IACSIT International Journal of Engineering and Technology (2010)

8. Wohlfarth, T., Ichise, R.: Semantic and Event-Based Approach for Link Prediction. In: Yamaguchi, T. (ed.) PAKM 2008. LNCS (LNAI), vol. 5345, pp. 50–61. Springer, Heidelberg (2008)
9. Pavlov, M., Ichise, R.: Finding experts by link prediction in co authorship networks. In: Proceeding of the 2nd International Workshop on Finding Expert on the Web with Semantics (2007)
10. Pallavi, R.N.: A Heuristic based technique for inferring links in Social Network. In: International Conference on Recent Advances in Engineering and Technology (ICRAET 2012), Hyderabad, India, April 29-30 (2012)
11. Katz, J.S.: Geographical proximity and scientific collaboration. Scientometrics (1994)
12. Melin, G., Persson, O.: Studying research collaboration using co- authorship. Scientometrics (1996)

P-Skip Graph: An Efficient Data Structure for Peer-to-Peer Network

Amrinderpreet Singh and Shalini Batra

Abstract Peer-to-peer networks display interesting characteristics of fast queries, updation, deletion, fault-tolerance etc., while lacking any central authority. Adjacency Matrix, Skip-Webs, Skip-Nets, Skip-List, Distributed Hash Table, and many more data structures form the candidature for peer-to-peer networks, of which, Skip-Graph (evolved version of skip-list) displays one of the best characteristics as it help to search and locate a node in a peer-to-peer network efficiently with time complexity being O(log n). However when a hotspot node is searched and queried again and again, the Skip-Graph does not learn or adapt to the situation and still searches traditionally with O(log n) complexity. In this paper we propose a new data structure P-skip graph, a modified version of Skip graph, which reduces the search time of a hot spot node drastically from initial time of O(log n). Results provided by Simulations of a skip graph-based Peer-to-Peer application demonstrate that the proposed approach can in fact effectively decrease the search time to O(1).

1 Introduction

Distributed peer to peer networks present a decentralized, distributed method of storing large data sets. Information is stored at the hosts and queries are performed by sending messages between the hosts as to identify the host(s) that store(s) the requested information. In other words, peer to peer networks are distributed systems without any central authority that are used for efficient location of shared resources. Some of the desirable features of peer-to-peer network are decentralization, scalability, fault-tolerance, self-stabilization, data-availability, load-balancing, efficient and complex query searching, *etc.*[1,2].

Amrinderpreet Singh
Samsung Engineering Lab (R&D), Noida

Shalini Batra
Thapar University, Patiala

© Springer International Publishing Switzerland 2015   43
R. Buyya and S.M. Thampi (eds.), *Intelligent Distributed Computing*,
Advances in Intelligent Systems and Computing 321, DOI: 10.1007/978-3-319-11227-5_5

Simplified P2P Architecture and Characteristics

The key feature of a peer-to-peer network is that this type of network lacks a central authority or a server to control the other participating nodes, reducing the probability of single point of failure. Furthermore, peer to peer system takes advantage of the computation power of each and every node and hence balances the load to all the nodes of the system. Since all the nodes are both client and server in P2P network, all nodes can provide and consume data. They also act as routers for forwarding the message from one node to the other in order to send the message to its final destination. Any node can initiate a connection as there is no centralized data source. Nodes or the systems have the computation capabilities which they can offer to the whole system while collaborating collectively. The nodes need not be in the same geographic area. The nodes may be distributed across the globe and may be connected to the peer network over the internet.

Some of the important characteristics of peer-to-peer network are:

- Clients are also servers and routers. Nodes contribute content, storage, memory, CPU, *etc.*
- Although nodes are autonomous (no administrative authority), they all share the same network.
- The network is dynamic i.e. the nodes enter or leave the network frequently.
- Nodes collaborate directly with each other (not through well-known servers).
- P2P needs to address some important issues like small nodes size, fault tolerance, fast queries and updates and support for ordered data before a data structure is selected for its implementation.

2   Literature Review

1960 saw the beginning of trivial distributed computing with the use of concurrent processes that communicate by message-passing, having its roots in operating system architectures[3]. Peer-to-peer distributed systems became popular in the Internet applications in a short period of time. Survey reveals a slew of desirable features for a peer-to-peer network such as decentralization, scalability, fault-tolerance, self-stabilization, data-availability, load-balancing, dynamic addition and deletion of nodes, efficient and complex query searching [4-7] incorporating geography in searches and exploiting spatial as well as temporal locality of the resources.

Recent systems like CAN [8,9], Chord[10,11], Pastry [12,13], Tapestry [2,14] and Viceroy [9,15] use a Distributed Hash Table (DHT) approach to overcome scalability problem. To ensure scalability they hash the key of a resource to determine which node it will be stored at and balance out the load on the nodes in the network. The main operation, in these systems, is to retrieve the identity of the node that stores the resource, from any other node in the network. To this end there is an overlay graph in which the location of the node and the resources is determined by the hashed values of their identities and keys respectively. Resource location using overlay graph is done in these systems by using different routing algorithms. Pastry and Tapestry use Plaxton's algorithm, which is based on hypercube routing [12-14]. The message is forwarded deterministically to a neighbour whose identifier is one digit closer to the target identifier. CAN partitions a d-dimensional space into zones that are owned by the nodes which store keys mapped to their zone. Routing is done by greedily forwarding messages to the neighbour closest to the target zone [8,9]. Chord maps nodes and resources to identities of m bits placed around a modulo 2^m identifier circle and does greedy routing to the farthest possible node stored in the routing table. Most of these systems use O(log n) space and time for routing and O($\log^2$ n) time for node insertion[10,11].

The initial systems such as Napster [16,17], Gnutella [18,19], and Freenet [20,21] did not support most of the features and clearly were unscalable either due to the use of central server (Napster) or due to high message complexity from performing searches by flooding the network (Gnutella).

In many optimization problems, such as p2p network linking and storage, as well as other applications, the input data, such as like node entry, failure, update, etc., are stochastic. In addition to the difficulties encountered in deterministic optimization problems, the stochastic problems introduce the additional challenge of dealing with uncertainties. Some of the systems discussed in this section have good search complexity, but are very sensitive to node failures and thus their performance degrade with random node failures.

To handle such problems, one needs to utilize probabilistic methods alongside optimization techniques. This led to the development of an advanced data structure Skip Graph [7,22], with an objective of providing tools to help design and control stochastic systems with the goal of optimizing performance. In Skip graphs main concern is dynamic allocation of memory and fast search results with minimum computational cycles. Interesting variants of skip graphs like skip webs [21,23], skip nets[7,22] and rainbow skip-graphs[5,7] have also been studied in the past due to the good results of skip graphs in p2p networks.

3 Proposed Data Structure: P-Skip Graph

A novel data structure, derived from traditional skip graph, P-Skip Graph is proposed in this work. In the proposed data structure a new level has been added to the top of the skip graph where initially all the nodes are in themselves a linked list with their right and left neighbours being empty. (This level is sometimes present in the traditional skip graph when all the nodes have entirely random but different membership vector).

Moreover traditionally the number of bits used in the membership vector are ceil (log n) -1. For e.g. if there are 8 nodes , levels = 3 , viz. 0,1,2, where level 0 has all the nodes connected in doubly linked list, level 1 has 2 linked lists whose least significant bit match. Level 2 has 4 lists where all the nodes have two LSB bits same. However in [7,22] and [5] authors have highlighted the fact that there can be another level i.e. level 3 but it would have all the nodes in their own linked lists, which would increase the number of levels to log n + 1 but the overall search complexity will grow at O(log n + 1) which is same as O(log n). But since the topmost level would contain no links to the neighbors', it is often ignored.

3.1   P-Skip Graph and the Next Level

It is assumed that pnode would refer to the node of a p-skip graph. A pnode consists of the following new fields along with the traditional fields:

Count_Vector: This vector keeps track of the number of times a particular key is searched by starting from this node. Every time a key is searched, the node updates its count_vector to reflect the number of times a particular key has been searched via this node.

Probability_Vector: Probability vector keeps track of the probability that a particular node will be searched in the near future by taking into account the previous searches through that node.

Top-Most P-Level: The topmost level, which is often left out in traditional skip graphs is now included as P-Level in the pnode. This P level is put into use by means of probability_vector. Since probability vector has all the calculated probabilities stored in it with respect to the keys that are searched, the P-level is used to create a link to the node with the highest probability.

Complexity of Skip Graph

When a key is searched in a traditional skip-graph (especially when the nodes are distributed across the internet) with high number of nodes, the search takes O(log n) time for the messages being passed between peers (which may be routed through the internet via standard routing protocols).

Let's assume that there are approx 65535 members in a p2p network and they share a skip-graph network for communication. The maximum number of possible levels would be log (65535) = 16. When a key is searched by a node, there are O(log n) messages passed to get the key i.e. 16 messages in the worst case.

Now consider a scenario that a particular resource has become very popular and is requested time and again from a particular node. Now, let's assume that a node receives 100 requests for the same resource. Every time the node receives a request, it sends the message via the same nodes, thereby congesting the network with 16 * 100 = 1600 messages.

Fig. 1 shows the Node Structure in P-Skip Graph where P-Level (indicated in red) is used. In the proposed structure, as the count of a particular key increases, the probability of the key being queried in future also increases. When the

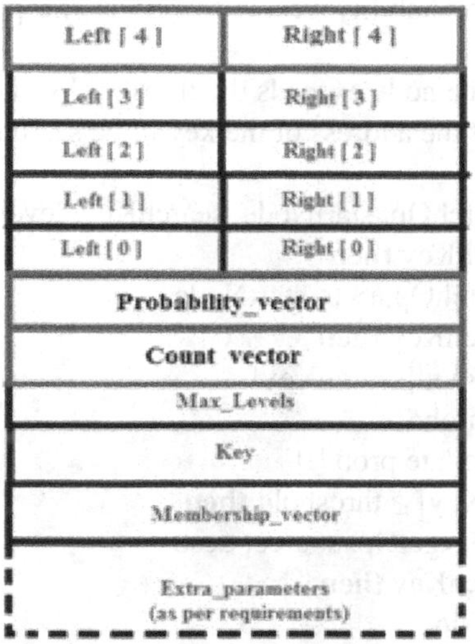

Left [4]	Right [4]
Left [3]	Right [3]
Left [2]	Right [2]
Left [1]	Right [1]
Left [0]	Right [0]
Probability_vector	
Count vector	
Max_Levels	
Key	
Membership_vector	
Extra_parameters (as per requirements)	

Fig. 1 Variables stored in a typical node of P-skip graph

probability of a particular node exceeds the threshold value (set up by each node of its own), the P-level of the node creates a link with the target key so as to directly address the target node and route the message via the internet rather than routing the message to the nodes of the skip graph and creating unnecessary chaos. Thus once the target node is included in the P-level of the node all subsequent queries can be done in $O(1)$ time complexity while giving the benefit to the network nodes to continue with their own operations thereby making the network congestion free and improving the effective search time.

3.2 P-Skip Graph Operations

The operations of the P-skip-graph are exactly the same as the traditional skip graph. There is no change in the existing algorithms. However, only the search algorithm needs to be modified for constructing the P-level. Since P-level is just another level in a node, only the maxlevel needs to be incremented by 1.

3.2.1 Modified Search Algorithm for P-Skip-Graph

The algorithm is exactly the same as traditional skip graph, with the only difference being that when the node receives the request to search a node key it performs the following steps:

- It checks for the searchKey i.e. it checks that the node to be queried is not itself.
- After checking that node is not the same one, it increments the count vector to reflect that a new query has been generated.

- It then updates the probability vector to reflect the possibility of any future query.
- If the probability of the node exceeds the threshold value, the node calls the getaddr algorithm to get the address of the key to be searched.

```
Upon receiving <searchOp, startNode, searchKey, level>
    if n.key = searchKey then
        send<searchOp,n> to startNode
    if n.key != searchKey then
        increment count[searchKey]
    for all i in the prob:
                update prob [i]
    if prob [searchKey] ≥ threshold then
            getaddr<getOp, source, dest>
      if n.key<searchKey then
    while level 0 ≥ do
            if (nR_level).key < searchKey then
                    send (searchOp, startNode, searchKey, level) to nR_level
        break
    else
            level←level-1
    else
        while level 0 ≥ do
            if (nR_level).key < searchKey then
        send (searchOp, startNode, searchKey, level) to nR_level
            break
    else level←level-1
            if level <0 then
                send <searchOp, n> to startNode
```

Algorithm 1. The modified search algorithm for the p-Skip Graph

3.2.2   Getaddr Algorithm for P-Skip Graph

When a node receives getOpt option, it performs the following operations:

It checks if the destination is the same as the node itself and if it is same it sends a reply to the source giving it the address of the node so that the source could directly address the message to the target without following the traditional approach.

If the destination is not the node itself, it checks the key of the destination and forwards it according to the normal search algorithm (level by level)

upon receiving <getOpt, source, dest, dest.addr >
if n.key == dest.key **then**
send <replyOpt, source, dest, dest.addr> to source
else
if n.key < dest.key **then**
 while level>0
  if nR_{level} != NULL **then**
     send<getOpt, source, dest > to nR_{level}
     break
    else level <- level-1
 else
    while level>0
     if nR_{level} != NULL **then**
send<getOpt, source, dest > to nR_{level}
      break
 else level <- level-1
if level <0 **then**
send <errorOp, source, dest> to source

Algorithm 2. GetAddr algorithm for the p-Skip Graph

3.2.3 Reply Algorithm for the P-Skip Graph

When a node receives replyOpt option, it checks the target key and then adds the address of the target to its left or right p- level according to the order of traditional skip graph.

Upon receiving <replyOpt, source, dest, dest.addr>
  if dest.key<n.key **then**
   nLp = dest.sddr
  else nRp = dest.sddr

Algorithm 3. Reply algorithm for the p-Skip Graph

3.2.4 Search Operation in a P-Skip Graph

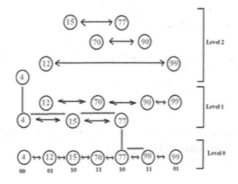

Fig. 2 (a)

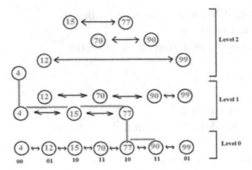

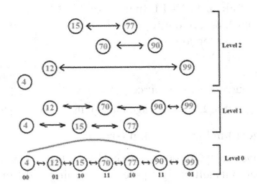

Fig. 2 (b)

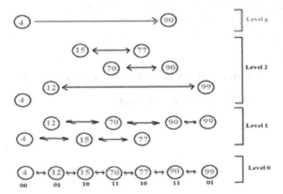

Fig. 2 (c)

Fig. 2 (d)

Fig. 2(a)-2(d) show how node with value 4 searches for a node with value 90 in a p-skip graph and establishes a direct link with it at level p

4   Implementation Details

The implementation of the p-skip graph in p2p network has been done in Linux environment using python (because of its rich socket library) and bash (for automating the process of file creation using bash scripts and utilities like sed, cut and bc). The nodes of the peer-to-peer network are created through files and script. However, to know what messages are being transmitted between the nodes netcat has been used to send and receive udp packets. All the nodes of the peer-to-peer network are executed on the same machine with different port numbers.

As explained in the Algorithm 1, p-skip graph takes into account the probability that a particular node would be searched frequently and includes the link to that node in the p-level of the source node. This makes a direct link to the target and hence reduces the further time complexity of the successive search queries for the same node. Fig. 3 shows the result of the search operation performed on skip graph and Fig. 4 shows the result of the search operation performed on p-skip-graph.

```
search 32969 55567
FOUND 55567 32969 55567 5
search 32969 55567
FOUND 55567 32969 55567 5
search 32969 55567
FOUND 55567 32969 55567 5
search 32969 55567
FOUND 55567 32969 55567 5
search 32969 55567
FOUND 55567 32969 55567 5
search 32969 55567
FOUND 55567 32969 55567 5
```

Fig. 3 Search Operation in Traditional Skip Graph Network depicting that even though the same query is being performed for the same target from the same source, search is always carried out from scratch

```
search 32969 55567
FOUND 55567 32969 55567 5
search 32969 55567
FOUND 55567 32969 55567 5
search 32969 55567
FOUND 55567 32969 55567 5
search 32969 55567
FOUND 55567 32969 55567 1
search 32969 55567
FOUND 55567 32969 55567 1
search 32969 55567
FOUND 55567 32969 55567 1
```

Fig. 4 Search Operation in p-Skip Graph Network, depicting that the number of hops decreases to 1 when the target gets added to the p-level of the source

The p-skip graph leverages the topmost level of the nodes left and right arrays and fills them probabilistically with appropriate link description. This search result message contains an additional field to calculate the number of hops the message travelled to reach the target (Fig. 3 and Fig. 4). Fig. 4 indicates that the bottleneck created by searching the nodes from scratch is removed by p-skip graph.

Fig. 5 show the search results achieved with trivial Skip-graph and P-skip graph with threshold value of 0.255. Threshold value can be fixed depending on the network's requirements and applications being considered.

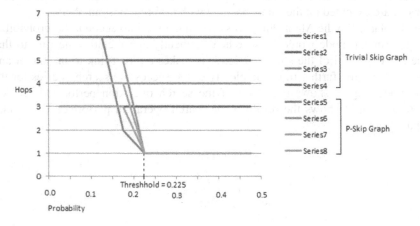

Fig. 5 Trivial Skip-Graph vs P-Skip Graph search result

5   Conclusion and Future Scope

P-Skip graph addresses the issue of reducing search time in peer-to-peer network by utilizing an extra level of the skip graph. Although extra level would increase the overhead by making the effective number of levels to be searched as log n +1 and the search complexity to be $O(\log (n) + 1)$ but essentially it is again equivalent to $O(\log n)$. Hence there is no degradation of traditional search query. Moreover, p^{th}-level introduced in P-skip graph contains a link to the most frequently searched target, reducing the search time from the node containing the p-level pointer to the target to $O(1)$. Further, since the count and the probability calculation is done every time a new search request is received, the time complexity of incrementing the count and the total count to find the probability is also of the order of $O(1)$.

The experiment has been performed locally on the same host machine with nodes running as different process on different ports. Thus the factors like internet congestion, message jitter have not been considered. The experiment, however, can be performed in the real world scenario with just a few minor changes in the python code to adjust to the real time peer-to-peer network over the internet. Due to the hardware limitations, the experiment has been performed with 16 peer-to-peer nodes. However with the availability of different host computers for running single node per host, the experiment could be extended to any number of nodes.

References

[1] Coulouris, G., Dollimore, J., Kindberg, T., Blair, G.: Distributed Systems: Concepts and Design, 5th edn. Addison-Wesley (2011)

[2] Schollmeier, R.: A Definition of Peer-to-Peer Networking for the Classification of Peer-to-Peer Architectures and Applications. In: Proceedings of the First International Conference on Peer-to-Peer Computing. IEEE (2002)

[3] Andrews, G.R.: Foundations of Multithreaded, Parallel, and Distributed Programming. Addison–Wesley (2000)

[4] Net History (2004),
    http://www.nethistory.info/History/email.html

[5] Goodrich, M.T., Nelson, M.J., Sun, J.Z.: The Rainbow Skip Graph: A Fault-Tolerant Constant-Degree Distributed Data Structure. In: SODA 2006 Proceedings of the Seventeenth Annual ACM-SIAM Symposium on Discrete Algorithm, pp. 384–393 (2006)

[6] Herlihy, M.P., Shavit, N.N.: The Art of Multiprocessor Programming. Morgan Kaufmann (2008)

[7] Aspnes, J., Shah, G.: Skip Graphs. Department of Computer Science, Yale University (2006), http://www.cs.yale.edu/SkipGraphs

[8] Ratnasamy, S., Francis, P., Handley, M., Carp, R., Shenker, S.: A Scalable content addressable network. In: Proceedings of the ACM SIGCOMM, pp. 161–170 (2001)

[9] Linial, N.: Locality in distributed graph algorithms. SIAM Journal on Computing, 193–201 (1992)

[10] Stoica, I., Morris, R., Karger, D., Kaashoek, F., Balakrishna, H.: Chord: A scalable peer-to-peer lookup service for internet applications. In: Proceedings of the SIGCOMM, pp. 149–160 (2001)

[11] Papadimitriou, C.H.: Computational Complexity. Addison–Wesley (1994)

[12] Andrews, G.R.: Foundations of Multithreaded. Parallel, and Distributed Programming. Addison–Wesley (2000)

[13] Rowstron, A., Druschel, P.: Pastry: Scalable, Distributed object location and routing for large scale peer-to-peer systems. In: Guerraoui, R. (ed.) Middleware 2001. LNCS, vol. 2218, pp. 329–350. Springer, Heidelberg (2001)

[14] Joseph, D., Kubiatowikz, J., Rao, S., Zhao, B.Y.: Tapestry: An infrastructure for fault tolerant wide area location and routing. Technical Report UCB/CSD-01-1141, University of California, Berkeley (April 2001)

[15] Malkhi, D., Naorand, M., Ratajczk, D.: Viceroy: A Scalable and dynamic emulation of the butterfly. 21st ACM Symposium on Principles of Distributed Computing, 183–192 (2002)

[16] Knighten, B.: Peer to Peer networking (2000),
    http://www.peer-to-peerwg.org/sep_docs/collateral/
    PtP_IDF_Rev.11.pd

[17] Napster Messages (2000),
    http://opennap.sourceforge.net/napster.txt

[18] Mayland, B.: Gnutella Protocols (2000),
    http://capnbry.dyndns.org/gnutella/protocol.php

[19] Hopfield, J.J.: Neural networks and physical systems with emergent collective computational abilities. Proc. Natl. Acad. Sci. USA 79, 2554–2558 (1982)
[20] Freenet Project (2000), https://freenetproject.org/
[21] Kaufmann, M., Pawel, P., Wellner Jr., R.: Grid Computing: The Savvy Manager's Guide (2006)
[22] Stone, E., Czerniak, T., Ryan, C., McAdoo, R.: Peer to Peer routing (2002), http://ntrg.cs.tcd.ie/undergrad/4ba2.05/group6/index.html
[23] Wilder, U., Shah, G.: Skip lists: a probabilistic alternative to balanced tree. Communications of the ACM (6), 668–676 (1990)

Localization in Wireless Sensor Networks with Ranging Error

Puneet Gour and Anil Sarje

Abstract. In wireless sensor networks (WSNs), localization is very important because there are many applications which depend on the location of a sensor node. Among various WSNs localization techniques, ranging methods based on received signal strength (RSS) are most popular because of simplicity and no additional hardware requirement. However, RSS ranging suffers from various environmental conditions and it can give an erroneous range for positioning of a node. Hence, it is necessary to deal with ranging error efficiently. In order to efficiently find the position of a node in the presence of RSS ranging error, we introduce a novel localization technique in this paper. To apply our positioning concept, our method selects three most suitable reference nodes according to RSS and geometry of reference nodes. We compare our simulation results with, localization with dynamic circle expanding mechanism (LoDCE), which clearly show that our method outperforms it.

Keywords: Localization, wireless sensor networks, trilateration, positioning, RSS.

1 Introduction

Wireless sensor networks have become more popular nowadays because of their numerous applications. There are many applications of WSNs, which are based on the location of the sensor node like target tracking, traffic monitoring, health and home applications, etc. Apart from these applications there are also some network services which use location of sensor nodes, like geographical routing, network deployment and intrusion detection systems. Hence localization is an active research area in WSNs. Localization can be defined as process of finding the geographical position of a sensor node within the network.

Puneet Gour · Anil Sarje
Department of Computer Science and Engineering
Indian Institute of Technology, Roorkee, 247667, India
e-mail: puneetgour10@gmail.com, sarjefec@iitr.ac.in

© Springer International Publishing Switzerland 2015   55
R. Buyya and S.M. Thampi (eds.), *Intelligent Distributed Computing,*
Advances in Intelligent Systems and Computing 321, DOI: 10.1007/978-3-319-11227-5_6

Global positioning system (GPS) is a very popular technique for localization, but it is not well suited for wireless sensor networks. Actually in WSNs a node has limited, battery power, computational power, memory, cost, etc. and due to these limitations GPS is not appropriate in WSNs. GPS also face problems in indoor environment. Hence, for solving these issues localization schemes have been introduced in wireless sensor networks. In WSNs localization, we use a few reference nodes known as "anchor" that have prior knowledge of their position (when deployed manually) or have the capability to find their position by themselves using GPS. All anchor nodes broadcast their position by a broadcast signal, which is called beacon signal. If any sensor node wants to find its position, it receives beacon signals and based on the position of anchors, it finds its position by a localization algorithm.

Most of the localization algorithms either fall into range based category or range free category. Range based methods require ranging information (distance) between anchors and the unknown node for localization. Hence, except RSS ranging, they require some additional hardware for doing this task. In case of range free methods, ranging information is not used and hence any additional hardware is not required. So range based methods are good in terms of accuracy, but not well in terms of cost, power, etc. whereas range free methods are useful when a sensor node has limited resources and coarse accuracy is acceptable.

Since the RSS ranging method does not require any additional hardware, we can say that it is suitable with the limitations of WSNs and can also give good accuracy. Our method also uses RSS based ranging technique to find the position of a sensor node. Whenever any node receives signal from an anchor then with the help of received signal strength it can estimate its distance from the anchor. Although in practical scenarios the signal strength is sensitive to noise, interference, reflection and other environmental conditions [1] and therefore these factors affect distance estimation accuracy. Hence it is very necessary to efficiently deal with distance estimation error to find the position of a sensor node. In, this paper we introduce a method which efficiently finds the position of a sensor node with distance estimation error.

2 Related Work

There are various methods for WSNs localization which can be classified based on ranging requirement between anchors and unknown node. Most of these methods either fall into range based category or range free category. Generally range based methods have two phases; first is ranging phase (distance estimation between anchors and unknown node) and the second one is position computation. Range based methods estimate distance from anchors in ranging phase and based on the estimated distances they apply second phase to find the position of a node. To estimate the distance some ranging techniques used are Time of arrival (ToA) [2] [3], Time Difference of Arrival (TDoA) [2] [4], Angle of Arrival (AoA) [2] [5], Received signal strength (RSS) [1] [2].

To estimate the distance between an unknown node and the anchors, ToA uses the propagation time of a signal which comes from anchors. But to find actual propagation time, nodes should have synchronized clock. To avoid synchronization requirement TDoA uses two types of signals and calculate difference between the arrival times of the two signals and based on this it estimates distance. But for the second type of signal it requires another transmitter and receiver. In AoA techniques some angle measurement hardware (like antenna) used for finding the direction of received signal and based on the directions of some signals a node calculates its position. Hence, all of the above-mentioned techniques require some additional hardware for ranging measurements. Due to additional hardware these techniques are not well suitable in terms of cost, battery power, size of node, etc. Another ranging technique which does not require any additional hardware is based on received signal strength. In RSS based technique a node estimates distance based on received signal strength. Range free methods do not require ranging information and based on positions of anchors they approximate the position of a node without requirement of any additional hardware. Some range free techniques are Approximate Point in Triangulation (APIT) [6], Range-free localization with the radical line [7] and DV-hop [8]. Range free methods do not give high accuracy as compared to range based methods, but they are effective in terms of cost, battery power and size of sensor nodes.

Range based methods are good in terms of accuracy as compared to range free methods, but face some other issues regarding the constraints of a node . Since signal strength can be measured by a node without the use of any additional hardware only RSS based ranging method removes the shortcomings of other range based methods, although RSS based ranging measurement is more sensitive to environmental situations as compare to other ranging techniques. The RSS ranging technique uses a theoretical or empirical model to translate the received signal strength information into the distance between sender and receiver. Once an unknown node finds its distance from some anchors (at least three), it calculates its position by positioning phase. For positioning it use some techniques like trilateration [2], bounding box [2] etc. Trilateration is one of the most popular methods. The basic principle of trilateration is that if we know the distances of an unknown point from at least three known points, we can find out the position of that unknown point. In reality the RSS ranging measurement is highly sensitive to noise, interference and reflection [1]. All of these have substantial impact on signal amplitude, and due to this it is practically very difficult to measure exact distance between anchors and unknown node. So with distance estimation error normal trilateration is not possible in all cases. There are various scenarios in which trilateration is not possible, these scenarios are described in the 3.2.2 section of this paper. It is necessary to deal with these cases efficiently. There are few mechanisms to deal with certain cases in [1], [9] and [10]. In this paper a novel mechanism has been introduced to find the position of an unknown node in all scenarios that may occur due to distance estimation error. Our method also focuses on selection of three anchors for positioning, when more than three anchors are available, because it influences the localization accuracy.

3   Basic Concept of Localization

First, all anchors broadcast their position periodically. If any unknown sensor node wants to find its position, it receives broadcast signal (beacon signal) from all anchors which are in its surrounding and then applies positioning technique to find its location. Fig. 1 shows a flow diagram which represents basic steps, performed by an unknown node.

After receiving beacon signals, there are basically two phases performed by an unknown sensor node to find its position.

3.1   Distance Estimation Phase

When a node receives beacon signal from an anchor, it receives it with a particular received signal strength (RSS) and based on this RSS, it calculates its approximate distance from anchor. To estimate the distance from anchors, unknown node can use any RSS-Distance model [1], [9], [11] depending on practical situations. To provide flexibilities in terms of environments and other issues, related to distance estimation phase, we took a simplest distance error model, which is also used by [6] for simulation of localization method, as follows:

$$EstDist = RealDist \pm random\ (e) \times RealDist \tag{1}$$

Where random (e) generates a random value between 0 and e, value of e is between 0 and 1 according to the percentage error in distance.

To compare our method we have also simulated a RSS-Distance model, which is used by LoDCE [1].

3.2   Positioning Phase

Once distance estimation phase is completed, a node use second phase for finding its position with distance estimation error. When RSS is used for ranging, the estimated distance is highly affected by noise. A very hot research area nowadays is how to deal with this ranging error efficiently such that a node can find its position with minimum possible error. So the main work of this paper is based on the positioning phase. To understand what the actual mechanism of positioning is, we start from the basic scenario of trilateration.

3.2.1   Trilateration

The basic principle of trilateration is that if we know the distance of an unknown point from three known points, then we can find out the position of unknown point. If we take a known point as center and the distance between known point and unknown point as radius, and draw three circles corresponding to three known points, then the intersection point of these circles would be the required point [2]. Fig. 2 shows trilateration scenario.

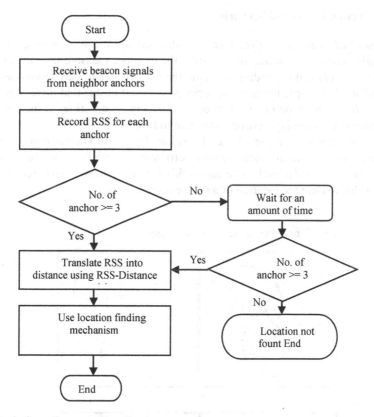

Fig. 1 Basic flow diagram of localization

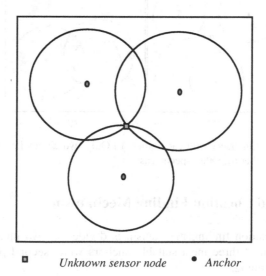

■ *Unknown sensor node* ● *Anchor*

Fig. 2 Trilateration Scenario

3.2.2   Actual Empirical Scenarios

The described scenario in Fig. 2 is very theoretical, and it is very difficult to practically achieve this scenario. In reality, the distance measurement is not much accurate [1], [11] and depending upon the inaccuracies there can be various cases [9], in which the simple trilateration is not possible. It means that there may be no common intersection point of all three circles. Fig. 3 describes some scenarios which may occur in noisy environments, due to ranging errors.

Any one scenario of Fig. 3 can be occur for an unknown node. So it is necessary to deal with all such scenarios efficiently. Very few methods consider all these scenarios effectively whereas in this paper we have considered all these scenarios for finding the location of a sensor node.

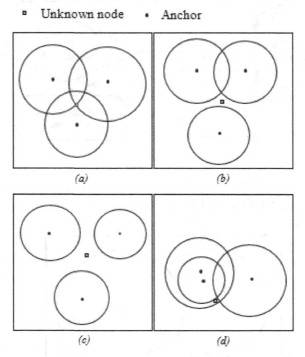

Fig. 3 (a) All circles intersect with each other (b) Only two circles intersect (c) No circle intersect with each other (d) Other special case

4   Proposed Location Finding Mechanism

Our proposed location finding mechanism is divided in two phases. First phase involves selection of three most suitable anchors while second phase uses these anchors for positioning.

4.1   Selection of Three Most Suitable Anchors for Positioning

Trilateration is one of the most popular methods for positioning due to its simplicity and less computational cost. But when the number of anchors surrounding an unknown node is more than three, a problem arises that which three anchors should be considered for trilateration. In case of ranging error the accuracy of trilateration depends on the geometry of anchors (relative positions of anchors) [12], [13], [14] and the size of the triangle made by anchors (RSS value of each anchor) [12], [14]. To select three anchors for positioning, our method considers both factors. To find out the geometry of three anchors by an unknown node, [12], [14] provide few techniques. In order to find appropriate geometry, we can say that an unknown node must fall inside the triangle which is formed by three anchors [14]. Some techniques to check this are given in [6], [14]. Due to various overheads we cannot use APIT [6] in our method. Based on the RSS based distance, we can predict that a point would be in a triangle or not, but RSS based distance is erroneous. To find the geometry of anchors we use a very simple technique which can also bear RSS distance error up to certain limits. In our technique we check the geometry of each triangle which can be created by neighbor anchors. In practice the number of neighbor anchors is small, so it will take a constant amount of time. In case the number of anchors is large, our method first selects some anchors (depends on particular situations) based on RSS value (anchor which gives large RSS value is preferable), so that our algorithm could take a constant amount of time for such task. In our technique we consider that an unknown node would be in a triangle (created by three anchors), if distances of an unknown point from all three vertices of the triangle are less than respective altitudes from those vertices. If this condition is false, we will not consider a particular geometry as acceptable geometry. Fig. 4 (a) represents a geometry where given condition is true and Fig. 4 (b) represents a geometry where geometry condition is false. Our given condition selects some particular geometries which would be more appropriate for trilateration. It ignores geometries in which unknown node is far from centroid of the triangle even it may be in triangle, because at that position chances of wrong geometry prediction (due to RSS error) would be higher. After selecting some good geometries, we choose a triangle which has the smallest sum of distances from all vertices of the triangle to unknown point. It is because anchors having less distance from unknown node

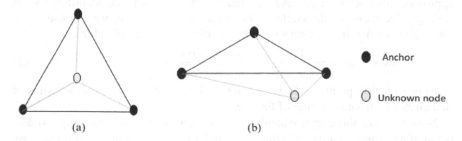

(a)   (b)

Fig. 4 (a) Geometry condition satisfied (b) Geometry condition not satisfied

have less error probability. If there is no combination of anchors, which satisfies the appropriate geometry condition, then we choose three anchors according to their RSS value. Fig. 6 shows Pseudo-code for selecting three anchors.

4.2   Positioning Based on Three Anchors

Fig. 3 shows that how various types of difficulties can be arise for trilateration. For an approximation of the position of a sensor node, our method uses a basic principle which says that the circle created based on high RSS value should give high confidence as compared to the circle created based on low RSS value [1], [10]. We can say that probably there may be less error in the circle of low radius and high error in the circle of high radius. Finally, we can say that probably the sensor node would be closer to the circumference of less radius circle as compared to high radius circle. To use the above-mentioned concept, our method assigns different weights to different circles. If any point is the intersection point of two circles, then average of the weight of both circles would be the weight of that point.

When a node has three circles having radius based on corresponding RSS value of three anchors, then there will be three different pair of circles. In our method we get an approximation point from each pair of circles, then calculate the final position based on all approximation points. Fig. 6 shows the pseudo-code and Fig. 7 shows the flow diagram for this mechanism. We consider each pair of circles one by one and check whether two circles intersect with each other or not. Depending on this, there are two mechanisms as follow:

When two circles intersect with each other

In this case there will be either one intersection point or at most two intersection points. If there is only one intersection point then we select this point as an approximation point for a pair. If there are two intersection points, then the sensor node may be near to either one of these two points. To resolve this ambiguity we consider the third circle, which is based on the RSS value of third anchor. Now we select one out of the two intersection points which would be closer to the circumference of the third circle. In Fig. 5 circles for anchors A1 and A2, intersect with each other and there is two intersection point, then an approximation point for that pair of anchors is selected according to third anchor A3.

When two circles do not intersect with each other

In this case first we consider a line between the centers of two circles and based on the value of RSS of each anchor, we map a point on that line. Fig. 4 shows approximation points for pair A2, A3 and A1, A3, which taken according to this concept. The radius of the circle is also a function of RSS, so we can use (2) to find a point on that line, which would be according to weight of radius.

$$\text{Division of line } = \frac{\text{Distance between centers of circles}}{\text{radius of first circle} + \text{radius of second circle}} \tag{2}$$

Approximation point would be a point, which is (division of line × radius of first circle) far from the center of first circle.

Now we have three approximation points, each point represents approximate position according to a pair of circle. For finding the final position of a node, we assign weight 'w' to each approximate point based on (3) [1].

$$w = \frac{1}{\text{radius of first circle} \times \text{radius of second circle}} \qquad (3)$$

After assigning the weight, we calculate the coordinates of final position as the weighted average of three points by (4), (5).

$$x = \frac{w_1 x_1 + w_2 x_2 + w_3 x_3}{w_1 + w_2 + w_3} \qquad (4)$$

$$y = \frac{w_1 y_1 + w_2 y_2 + w_3 y_3}{w_1 + w_2 + w_3} \qquad (5)$$

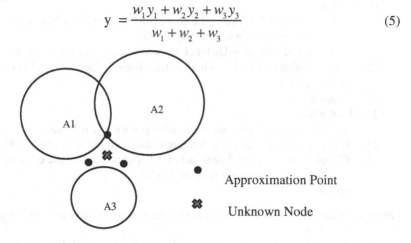

A1   A2   A3

●   Approximation Point

✖   Unknown Node

Fig. 5 Approximation points for pairs of circles

Algorithm for location finding mechanism
When any unknown node wants to find its position and if it is in the range of at least three anchors then it applies this algorithm.

Input: The estimated distance from neighbor anchors.
Output: Estimated position of unknown node.
1. **If** number of neighbor anchors >3
2. Select three anchors with the help of function **Three_Anchor_Selection** ()
3. **End if**
4. Consider three circles C1, C2, C3 having center at the coordinates of anchors, with corresponding radius r1, r2, and r3 which are corresponding distance from the node to anchors.
5. **If** C1, C2, C3 intersect to each other at a common point
//Trilateration
6. **Return** the intersection point as coordinates of the node
7. **Else**
8. **While** Each pair of circles does not have an approximation point do
    //step 8   to 24 for each pair of circles e.g. C1 and C2
9.   **If** C1 and C2 intersect to each other
10.   **If** C1 and C2 intersect at a common point I1
11.   Approximation point A1= I1

12.   **Else**
           // Select an intersection point I1 or I2, which is closer to
           circumference of third circle C3.
13.   **If** I1 is closer to the circumference of C3
14.   A1=I1;
15.   **Else**
16.   A1=I2;
17.   **End if**
18.   **End if**
19.   **Else**
20.   Create a line L1 between the center of C1 and C2. // Map a point on
       this   line according to the proper weight.
21.   Division of line L1 = Distance between centers of C1 and C2/ (r1+r2)
22.   A1= a point on line L1, which is ((division of line L1) × r1) far from
       center of C1
23.   **End if**
24. **End while**
       // Now we have three approximation points A1, A2, A3.
25.   Calculate the weight of each approximation point according to (3).
26.   Calculate the weighted centroid of A1, A2 and A3 as the estimated
       coordinates of unknown node by (4), (5).
27. **End if**

Three_Anchor_Selection () // for selecting three anchors when neighbor anchors
are more than three.
Input: Coordinates of each neighbor anchor and correspond distance from
unknown node.
Output: Coordinates of three anchors.
   1. Prev_Total_Distance = Infinite
   2. **While** each triangle of a combination of three anchors does not
       check out **do**
   3. **If** all distances of the unknown point from three vertices of the
       triangle   are less than or equal to corresponding altitude
   4. Total_Disatance = sum of distances of unknown point from three
       vertices of triangle
   5. **If** Total_Distance < Prev_Total_Distance
   6. Update three anchors for positioning
   7. **End if**
   8. **End if**
   9. **End while**
   10. **If** Prev_Total_Distance==Infinite
   11. Sort all anchors based on RSS value
   12. Select three anchors for which unknown node has smallest RSS
       value
   13. **End if**
   14. **Return** Coordinates of three anchors

Fig. 6 The Pseudo-code of the location finding mechanism

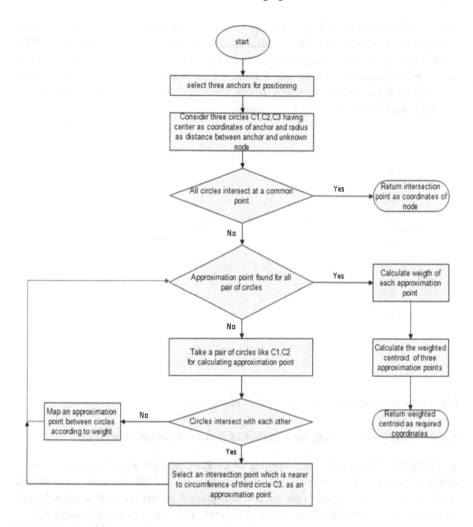

Fig. 7 Location finding mechanism

5 Performance Evaluation

To evaluate performance of our method, we have simulated our algorithm in MATLAB. We did the evaluation in two phases first is independent analysis and second phase is comparative analysis of our method.

5.1 Analysis of Proposed Method

To evaluate performance of our method we have done simulation of our algorithm on 10×10 meter square area with the uniform deployment of 100 unknown sensor nodes. To provide flexibility with environmental situations we kept our method independent from the first phase (Distance Estimation Phase) of localization. We used a simple Distance estimation model given in (1) for our simulation, where noise error was up to 20% of real distance. To evaluate proper localization error each node finds its position ten times, then the average of all iterations is considered as localization error for a node.

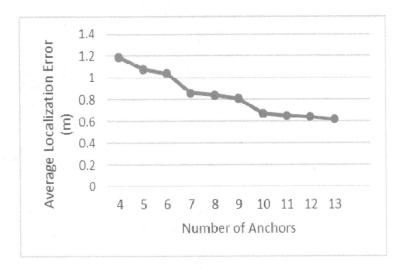

Fig. 8 Average localization error vs. Number of anchors

5.1.1   Effect of Varying the Number of Anchors (Reference Nodes)

To study the effect of different number of anchors we performed simulations with different number of anchors. Communication range of all anchors were 10 m. Each time we incremented one anchor and placed all anchors on some selected positions in the network area. When the number of anchors is increasing for an unknown node, unknown node can select a more appropriate combination of three anchors according to the geometry and the RSS value of the anchors. As we described in section 4.1, geometry and RSS values for a combination of anchors can influence on localization error. As soon as we increment number of anchors, the average localization error reduces. Fig. 8 shows the impact of number of anchors on localization error.

5.1.2   Effect of Communication Range of Anchors

When the communication range of anchors is large, then an unknown node may have some anchors at larger distance. Hence, the probability of noise error would be higher, resulting higher localization error. Although number of sensor nodes capable of finding their position will be increase. In some cases the localization error may be reduced because an unknown node can select an anchor which is far from it, but overall good according to its geometry. To study the impact of communication range, we have done different simulations with different communication range of anchors. Each time we incremented the distance of anchors from unknown nodes and the communication range of anchors. First, we took 10m×10m area and deployed 100 unknown nodes uniformly. Then we placed four anchors at each corner of the area and one anchor at the middle. Initially communication range of anchors was 10 m. After finding the average localization error in that case, now we shifted only four anchors which were at corner of network area. Now four anchors would be at four corners of a square having the same middle point as previous case and side 12 m, so now communication range will be 12 m. Then, unknown nodes again apply a fresh simulation for localization. Each time we shifted only corner's anchors and found the average localization error up to 28 m communication range. Fig. 9 shows the impact of communication range on localization error when the distance of anchors from unknown nodes were changed in above described manner.

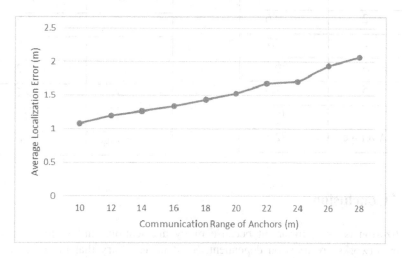

Fig. 9 Average localization error vs. Comm. range of anchors

5.2   Comparative Analysis

To compare our method with LoDCE [1], we have applied our method on same RSS-Distance model, which is used by LoDCE [1]. We have taken the same parameters which given in LoDCE [1] to find location estimation error under random deployment of sensor nodes. In our simulation, an unknown node finds its position thousand times and then we take the average error of all iterations for localization error of each node. We have run our simulation multiple times and found that each time our method outperforms LoDCE [1] (best case result is given in [1]). Each node has a different scenario for positioning because there was a random deployment, but each scenario has some conditions given in [1]. Hence, we have taken six random scenarios and given localization error in them. Results of LoDCE [1] and our proposed method are shown in Table 1 and Table 2 respectively. Parameters we have used for simulation, which are same as used by LoDCE [1] are as follow:

Network Area = 7.08 m × 10.6 m, Number of Anchors = 20, Unknown Nodes = 6
Maximum distance between an unknown node and anchor for positioning = 0.7 m
Minimum distance between two anchors for positioning = 0.53 m

Table 1 Localization Error (in cm) in LoDCE

Table 2 Localization Error (in cm) in Proposed Method

Unknown Node	Localization Error
1	8.4
2	29.6
3	22.7
4	56.2
5	11.8
6	6.1
Average	22.5

Unknown Node	Localization Error
1	10.9
2	6.2
3	7.8
4	8.2
5	9.1
6	11.9
Average	9.0

6 Conclusion

Localization is very important because many applications and various tasks in sensor networks are location dependent. So it is necessary that the accuracy of localization should be high as possible. Due to various constraints of sensor node we cannot use additional hardware in each sensor node in general. Hence, RSS ranging based solution is good for localization, but it is highly sensitive with various environmental conditions. It is necessary that a method could deal with ranging error efficiently. To overcome RSS ranging based issues for positioning, we have introduced a new method for localization, which deals with all cases that may occur due to ranging error, where normal trilateration is not possible. The simulation results of our method clearly show that our method outperforms other methods. Our method can be modified for three dimensional localization in future.

References

1. Jiang, J.-A., Zheng, X.-Y., Chen, Y.-F., Wang, C.-H., Chen, P.-T., Chuang, C.-L., Chen, C.-P.: A Distributed RSS-Based Localization Using a Dynamic Circle Expanding Mechanism. IEEE Sensors Journal 13(10), 3754–3766 (2013)
2. Boukerche, A., Oliveira, H.A.B., Nakamura, E.F., Loureiro, A.A.F.: Localization systems for wireless sensor networks. IEEE Wireless Communications 14(6), 6–12 (2007)
3. Lanzisera, S., Lin, D.T., Pister, K.S.J.: RF Time of Flight Ranging for Wireless Sensor Network Localization. In: 2006 International Workshop on Intelligent Solutions in Embedded Systems, June 30-July 30, pp. 1–12 (2006)
4. Savvides, A., Han, C.-C., Strivastava, M.B.: Dynamic fine-grained localization in ad-hoc networks of sensors. In: Proc. 7th ACM MobiCom, pp. 166–179 (July 2001)
5. Niculescu, D., Nath, B.: Ad hoc positioning system (APS) using AOA. In: Twenty-Second Annual Joint Conference of the IEEE Computer and Communications, INFOCOM 2003, March 30-April 3, vol. 3(3), pp. 1734–1743. IEEE Societies (April 2003)
6. Bachrach, J., Taylor, C.: Localization in sensor networks. In: Handbook of Sensor Networks. Wiley (2005)
7. Chen, H., Chan, Y.T., Poor, H.V., Sezaki, K.: Range-Free Localization with the Radical Line. In: 2010 IEEE International Conference on Communications (ICC), May 23-27, pp. 1–5 (2010)
8. Gui, L., Wei, A., Val, T.: A Two-Level Range-Free Localization Algorithm for Wireless Sensor Networks. In: 2010 6th International Conference on Wireless Communications Networking and Mobile Computing (WiCOM), September 23-25, pp. 1–4 (2010)
9. Karagiannis, M., Chatzigiannakis, I., Rolim, J.: Multilateration: Methods For Clustering Intersection Points for Wireless Sensor Networks Localization with Distance Estimation Error. arXiv preprint arXiv: 1203.3704 (2012)
10. Sharma, N.K.: A weighted center of mass based trilateration approach for locating wireless devices in indoor environment. In: Proc. 4th ACM International Workshop on Mobility Management and Wireless Access, pp. 112–115. ACM (2006)
11. Adewumi, O.G., Djouani, K., Kurien, A.M.: RSSI based indoor and outdoor distance estimation for localization in WSN. In: 2013 IEEE International Conference on Industrial Technology (ICIT), February 25-28, pp. 1534–1539 (2013)
12. Yang, Z., Liu, Y.: Quality of Trilateration: Confidence Based Iterative Localization. In: The 28th International Conference on Distributed Computing Systems, ICDCS 2008, June 17-20, pp. 446–453 (2008)
13. Sadaphal, V.P., Jain, B.: Localization accuracy and threshold network density for tracking sensor networks. In: 2005 IEEE International Conference on Personal Wireless Communications, ICPWC 2005, January 23-25, pp. 408–412 (2005)
14. Zhang, A., Ye, X., Hu, H.: Point in Triangle Testing Based Trilateration Localization Algorithm in Wireless Sensor Networks. KSII Transactions on Internet & Information Systems 6(10) (2012)

Dynamic Job Scheduling Using Ant Colony Optimization for Mobile Cloud Computing

Rathnakar Achary, V. Vityanathan, Pethur Raj, and S. Nagarajan

Abstract. Cloud computing has been considered as one of the important computing paradigm. Its main purpose is to share computing resources. With the current scenario there is no doubting the incredible impact that mobile technologies have had on both in scientific and commercial applications. The integration of emerging cloud computing concept and the potential mobile communication services is together considered as Mobile Cloud Computing (MCC). A prominent challenge by using mobile devices and the mobile cloud [1] is resource constraints of these handheld devices. The computational complexities in mobile devices compared to the desktop computers are due to its smaller screen size, less memory capacity, lower processing capacity and low battery backup. Due to these resource limitations most of the processing and data handlings are carried out in the cloud, which is known as software as a service (SaaS) cloud. The smart phones are used to access could resources by using the browser. Performance of this mobile cloud is impaired by the time varying characteristics such as, latency, jitter and bandwidth of the wireless channel. In this research we propose a modified task scheduling mechanism called Ant Colony Optimization (ACO) to address the issues related to the performance of mobile devices [5] when used in a cloud environment and Hadoop. However there are bottlenecks related to the existing task scheduling techniques in MCC model which uses the built in

Rathnakar Achary
Department of Computer Science, Alliance University, Bangalore, India
rathnakar.achary@alliance.edu.in

V. Vityanathan · S. Nagarajan
Department of Computer Science and Engineering, Sastra University, Thanjavur, India
e-mail: vvn@it.sastra.edu, nagarajan@cse.sastra.edu

Pethur Raj
Infrastructure Architect, IBM Global Cloud CoE, IBM, India
e-mail: peterindia@gmail.com

© Springer International Publishing Switzerland 2015   71
R. Buyya and S.M. Thampi (eds.), *Intelligent Distributed Computing*,
Advances in Intelligent Systems and Computing 321, DOI: 10.1007/978-3-319-11227-5_7

FIFO algorithm for large amount of tasks. The proposed Ant Colony Optimization algorithm improve the task scheduling process by dynamically scheduling the tasks and improve the throughput and quality of service (QoS) of MCC.

Keywords: Mobile Cloud Computing (MCC), Ant Colony Optimization (APO), Hadoop, Quality of Service (QoS).

1   Introduction

Due to the advancement in cloud computing many researchers have developed the mechanism to show, how smart phones can be used to form an independent cloud computing network [7]. When smart phones are used as a part of the cloud infrastructure, they have the similar functionalities in terms of processing power as that of server, also the benefits of using these devices over a network. The users of these wireless devices are facing many challenges. These are due to the limited resources such as processing capacity, memory and operating systems, considerably hinder the enhancement of QoS. The key challenges for MCC are;

- Higher network throughput required for the real time data transmission between the master and the slave.
- Lower delay characteristics of the network such as latency and jitter which cannot be tolerable by certain applications.
- Isolating the application functions across the cluster and the wireless device to optimally utilize the resources.
- In a cloud the Hadoop typically run on server. Data and file transmission from the client and the master requires a very high bandwidth of the order of gigabits per second, which is many times more than that of the bandwidth of the Wi-Fi links. This results the MapReducer [11] job on a wireless mobile cluster [3] would be expected to perform much worse than a traditional cluster.

Under the mobile cloud computing environment in regard to multi-user with small amount of task scheduling from master to slave nodes Hadoop based architecture is used. This has certain advantages and it also uses FIFO algorithm for scheduling the tasks. Being next generation MCC platform needs with large amount of different granularity concurrent tasks to be processed in both master and slave nodes. The static task scheduling like FIFO is not suitable to these types of applications. Dynamic task scheduling is one of the solutions to mitigate these problems. The purpose of this research is to implement a dynamic job scheduling technique using Ant Colony Optimization as an effective algorithm to schedule the tasks for mobile cloud computing infrastructure on Hadoop [1],[12]. We organized this paper as follows: In section 2 we explained the related work about mobile cloud computing (MCC), in section 3 an overview of the Mobile Cloud Computing (MCC), section 4 explore the MapReduce. We then detail our design and analysis Ant Colony Optimization algorithm in section 5. Section 6 gives the details of results and performance evaluation. The section 7 concludes.

2 Related Work

In [1] and [12] the author explained the objectives of Apache Hadoop. It was provided as a protocol, to list and evaluate a huge number of web contents. By default all the tasks scheduled by Hadoop, using FIFO queue. This mechanism is very in-efficient. The modified scheduler in Hadoop is Hadoop on Demand (HoD) to address this concern by providing a private MapReduce [11] scheduling system. This technique also failed because it desecrated the design features like data locality of the initial MapReduce scheduler and added an additional overhead. The authors in [2] and [6] implemented the techniques of ACO for parallel problem solving and computational intelligence. ACO is one of the efficient techniques with strong job distribution capability, scalability and parallelism. In the existing Hadoop cloud computing system, if the number of nodes entering into the cloud increases, it degrades the performance and also there is a probability of failure. We combined the advantages of ACO into MCC to achieve an efficient and scalable job scheduling mechanism.

3 Mobile Cloud Computing

MCC refers to an infrastructure. In which we expect that, the execution and data storage happen in the cloud. i.e, in MCC all these functions are transferred from the mobile device to high performing computing systems located in the cloud. These applications located in the centralized systems are then accessed over the wireless link using mobile devices. We have included the key benefits of MCC in the following section;

3.1 Extended Battery Life

Battery is one of the key resources in any mobile device. Computation technique is one of the mechanisms introduced with the purpose of migrate the large computation from resource limited mobiles devices to the cloud. This avoids the execution of long applications on the hand held devices and smart phones. Which minimizes the power consumption [10]. The cloud computing can save the energy significantly.

3.2 Improved Data Storage Capacity and Processing Power

Mobile devices [5] have limited storage capacity. Using MCC the users can store large amount of data on the cloud through wireless networks. Users can upload the data to the cloud and can access them from mobile devices. This can save considerable amount of storage space on their mobile devices.

3.3   *Improving Reliability*

Keeping user applications and data on the wireless cloud is an effective way of improving the systems reliability. These data and applications are maintained in number of computers in the cloud for backup. The multiple copies minimize the risks of data loss.

4   Mapereduce

MapReduce [11] address the programming challenges, large data processing and scheduling in a cloud environment. It was originally a Google technology developed for large scale data processing in a distributed computing environment. MapReduce is a simple programming concept that hide low level details from the programmer at runtime, tuned to the basic architecture. In this the user simple specify a Map function. It processes the data to generate a set of intermediate key/value pairs. A Reduce function that processes all intermediate values associated with the same intermediate key. The system consists of a client server environment with a master server and several wireless client terminals. Where the server performance the duties of monitoring and keeping track of the user data and applications. It also take the responsibility of job scheduling and assigning them to the respective clients. The clients retrieve the respective modules and data from the server and, then perform the execution on them. It then return the results to the server. One of the key issue in the implementation process of MapReduce on the mobile devise is that, in a regular cloud with servers, the failure rate is very low, and the latency of a signal transmitted between the nodes is relatively minimum. This may not be in the case of mobile devices. It means that the failure of any one node in the system declines the performance of the entire network exponentially. The MapReduce system will automatically handle the distribution of codes, distribution of data, executing the tasks in parallel, work scheduling and all other complicated and distribution systems issues.

In the MapReduce [11] mode as in fig.1, the map and reduce are the data processing functions. The parallel map tasks are run on input data which is partitioned in to fixed size blocks and yield intermediary output as a group of <*key, value*> pairs. These combinations are shuffled across different reduce tasks based on <*key, value*> pairs. Each reduce tasks accept only one key at a time and process data for that key and output the result as <*key, value*> pairs. Hadoop MapReduce architecture includes a job tracker (Master) and many Task Trackers (workers) [8], [9]. The job tracker receive the job as an input from user, divides it into map and reduce tasks, allocates these tasks to the task trackers, screens the progress of the task tracers, and finally when all the tasks are completed, the user will get the job accomplishment report. Each Task Tracker has a specific amount of maps and reduces tasks slots that define how many map and reduce tasks it can run at a time. In cloud computing large amount of different granularity concurrent tasks are to be processed. Using only static job scheduling technique like FIFO used in Hadoop is not suitable to this application. Dynamic job scheduling algorithm has great stochastic performance. The presence of master node mechanisms and overall task scheduling mechanism of Hadoop [12] can split the input file into many blocks and dispatch them to different slave nodes and implement data locality to avoid large scale of data shuttle and save lots of processing time and input/output time. Hadoop includes a built in FIFO algorithm which sequentially execute the tasks according to the default priority parameters and the arrival time. This results into a larger waiting time for small granularity tasks, when large granularity tasks are under execution. This disadvantage would turn out many resource fragments, underutilization and poor flexibility. The first solution to this problem was Hadoop on Demand (HoD) [12], which provides private MapReduce clusters over a large physical cluster. In an infrastructure based wireless cloud, the master node has higher performance than slave nodes. The objective of job scheduling is to dispatch parallel jobs to slave nodes according to scheduling policy and priority constraints to reduce total execution time and minimizing the computation cost, which intern improve the performance and the Quality of service (QoS).

5 Ant Colony Optimization

Ants always find a shortest path from their nest to the food with the help of pheromone. They uses pheromone for the communication among themselves. This guide them for the next movement. Based on the intensity of the pheromone deposited on different paths, ants decide the shortest distance to reach the destination. In the proposed algorithm routing information is provided in a two dimensional pheromone table $P^k{}_{ij}$. The value $P^k{}_{ij}$ entered in the table provides routing details starting from node i, to destination d, over neighbor j. This provide the pheromone value, which gives the quality of the route from source to destination, as well as statistical details about the path and possible virtual pheromone. The table also maintains information about the neighboring nodes, it has wired or wireless links too. In the beginning of the communication process the source node control its pheromone table. It is later used for finding the routing details related to the intended destination. If it does not starts a reactive route setup process, then it sends an ant packet out over the network. This determines the path for the destination and the packet used for the same is called as reactive forward ant. When it moves the intermediate nodes also receive a copy of this reactive forward ant. This process is done by unicasting in case the node has routing information's else the details are broadcasted. The details gathered by the reactive forward ant, during its visit to the destination is stored. The initial details provided by the reactive forward ant to reach the destination is later transformed into a reactive backward ant; this repeats the particular path, which was followed by the forward ant back to the source. On its way again it collects the details about the entire route. These details are used to update the routing tables. This is how the first route between the source and destination is established at computation of the reactive route setup processes.

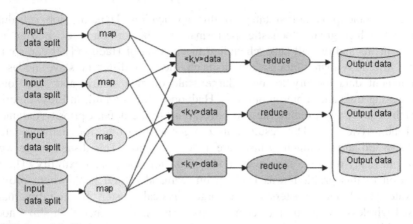

Fig. 1 Hadoop MapReduce

5.1 ACO Algorithm and Analysis

The task scheduling issue is addressed as follows. If there are n tasks, it needs to be dispatched from n wireless Nodes (Smart phones) to the cloud, where one smart phone takes one task at a time. So different dispatch plans have different execution cost and resource consumption. The proposed algorithm is to find the plan to allot the tasks smoothly to ensure availability and optimality of the resources. Under the same processor performance condition the task complexity is the key factor to influence processing time. Complex task require higher processing time than simple jobs. The different parameters considered for optimization and scheduling is discussed below.

The processing cost for a ith node to complete j^{th} task is;

Cost matrix: $C_{n*n} = \{C_{i*j} \mid C_{i*j} \in C_{n*n}, \ C_{i*j} \geq 0; \ i = 1.2.3...........n; \ j = 1,2,3...........n$

These values in the matrix are derived from the requested task complexity and processor performance. $T_{n*n} = \{T_{ij} \mid T_{ij} \in T_{n*n}$

These values in the matrix represents the pheromone of that the jth task was dispatched to i^{th} node. The matrix is initialized to constant matrix or 0 before scheduling.

Performance matrix; $V_{n*n} = \{V_{ij} \mid V_{ij} \in V_{n*n}$

These values in the matrix represents that the jth task was dispatched to i^{th} node. The matrix is initialized as $1/C_{j*j}$ that is to say this matrix is reverse ratio to cost i^{th} Task scheduling matrix: $R^{k}_{n*n} = \{R^{k}_{ij} \mid R^{k}ij \in R^{k}_{n*m} \quad i = 1,2,3...........n; \ j = 1,2,3...........n; \ k = 1,2,3...........n$.

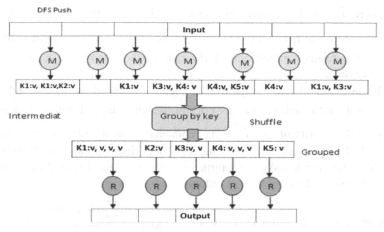

Fig. 2 MapReduce Work Flow

This represents the task scheduling plan that the kth ant implements task scheduling. The matrix is initialized as 0, and the value of elements in this matrix is 1 or 0. When, $R_{ij} = 1$, j^{th} task is dispatched to i^{th} node. $R_{ij} = 0$, j^{th} task is not dispatched to i^{th} node. The task dispatching matrix R_{n*n} represents how to dispatch the task to the slave node to complete all the submitted tasks with the minimum cost. If there are N ants to complete all the submitted tasks, ant's in one trip stands for one task dispatching procedure in which they need N walks which demonstrates dispatching of one task. These walks are tagged as 's'. When, all the ants complete one trip, and then it can be thought as one loop in completed. N_c - is the times of loops.

$$Task = \{task1, task2, \ldots\ldots\ldots.taskn\}$$
$$node = \{node1, node2, \ldots\ldots\ldots noden\}$$

Let us introduce an 'n' dimension vector $D^{N_n}{}_n$ which has elements $D^{N_n}{}_k$ that is $D^{N_n}{}_k \in D^{N_c}{}_n$ to stand for the kth ant's cost vector during the $N_c{}^{th}$ algorithm loop. The initial value of $D^{N_n}{}_n$ is 0. The key probability matrix $P^k{}_{n*m}$ which has elements $P^k{}_{ij}$ that is $P^k{}_{ij} \in P^k{}_{n*m}$ to stand for the probability of dispatching j^{th} jobs to i^{th} slave node, and $P^k{}_{ij}$ has relations to its pheromone matrix and its tasks performance matrix with their relationship is;

$$P^k{}_{ij} = \frac{T^{\alpha}{}_{ij} * V^{\beta}{}_{ij}}{\sum_{j=1}^{n} T^{\alpha}{}_{ij} * V^{\beta}{}_{ij} j}$$

Where T_{ij} - is the intensity of pheromone deposited in the route (i, j), V_{ij} - is the ant's visibility filed on the route (i, j), α and β - are the parameters which control the relative importance of the pheromone intensity compared to ant's visibility field.

The task arrival and completion moment are tagged as $at(i)$ and $et(i)$ respectively with each job has its priority, which is tagged with a Pr$iority(i,t\ )$ meaning the priority of task i at the moment t .The algorithm is described as in figure 3. If M tasks are arrived during a given period and there are N tasks which are submitted and k tasks are returned due to time out. Using these parameters there are some indexes of QoS as follows;

Average task Execution Time (AET);

$$AET = \frac{\{\sum_{i=1}^{n}(et(i) - at(t))\}}{N}$$

Task weighting Average execution time (WAET);

$$WAET = \frac{\{\sum_{i=1}^{n}(et(i) - at(t)) * Priority(i, et(i)))\}}{\sum_{i=1}^{n} Priority(i, et(i))}$$

The task's losing rate $ls = \frac{k}{M} *100\%$. The slave node work load ratio which is sampled at every period.

6 Result and Performance Evaluation

To verify the efficiency of our ACO algorithm we simulated the performance and other QoS parameters against Hadoop's built in FIFO on MCC environment [12]. The master node is a high performing computing device with Linux operating system as the platform and the slave nodes are the smart phones with Android operating systems. The performance is validated for Hadoop FIFO scheduling and by the proposed dynamic scheduling technique using ACO algorithm for scheduling the tasks. The static and dynamic scheduling performance comparison is done by simulating the system using NS2. The simulation system includes 10 wireless nodes and submitted an equal amount of jobs for both Hadoop FIFO algorithm and ACO based dynamic scheduling mechanism. Table 1 represents in ACO based dynamic scheduling the total execution time and loop time is approximately half of that for the static scheduling. The default FIFO algorithm in Hadoop results a best performance for small granularity jobs to achieve local optimization. For large amount of jobs and to attain global optimization, a better performance can be achieved by using dynamic scheduling using ACO algorithm.

```
Algorithm :Dynamic job scheduling
1. Initialize  N_c =1,Task, Node
2. Do
3.    for(K =1;k ≤n;k++)  Begine
4.     for(s =1; s≤n; s++)  Begine
5.      If
6.        the task is tagged Priority(i ,t)
7.        Randomely  select the node i to compute
8.        the j^th task at a Probabilit y P^k_ij
9.        Tag it as P^k_ijmax
10.       Set R^k_ij =1
11.       Delete the node i from nodeset
12.       Delete the taskj fromTaskse t
13.       set cost vector D^Nc_k = D^Nc_k +C_ij
14.    else
15.       wait for task jPriority ≥ Priority(i ,t)
16. End
17. End
18.   for(k =1; k ≤n; k++) Begine
19. Pheromone  update :
20. Increment  in pheromone  ÄT and ants pheromone
21. is gained to cost  inversly
22.    ÄT= Σ_{k=1}^{n} Q/DNCK and according  to
23.       the k^th ant R^k_n*n matrix
24.       Is R^k_ij =1 and set T_ij = T_ij + ÄT
25. End
26. set volatile  parameter  0<p<1 to
27.    limit the infinite  increment  of
28.    T = T*(1-p)
29.    Find the minimum  element
30.    D^Nc min among cost vector D^Nc_n
31. If D^Nc min < D^Nc min
32.    Nc = Nc + 1
33.    until Nc ≥Nc max
```

Fig. 3 ACO Algorithm

Table 1 Experimental Result of Job Scheduling

Trials	FIFO Scheduling		ACO based dynamic scheduling	
	Sum of loops	Sum of execution time	Loop time	Execution time
1	19	31ms	8	16ms
2	17	29ms	9	14ms
3	15	26ms	7	13ms

Fig. 4, represents the percentage parallelism (P) changes between the wireless clients and the servers. The graphs are plotted by considering the differences in the slop of the processing time and the task time versus the number of nodes connected. Fig. (5) and (6) represents the variations of data transfer time with respect to the size of the packets, of small and large jobs. The transfers of the small granular jobs are more costly in terms of number of bytes per second, particularly for the wireless links, because of the additional information provided for small packets becomes an overhead to establish the connection.

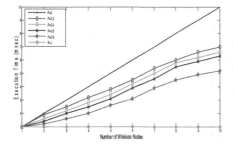 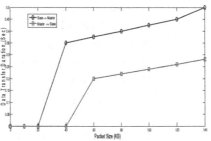

Fig. 4 Varying Execution Time with Percetage Parallelism

Fig. 5 Network Trnsfer Time for Small Granularity Jobs

Fig. 7 shows the average execution time (AET) for the two algorithms is approximately equal with less number of jobs. As the number of jobs inflow to the system increases the graph representing ACO indicates the enhancement in execution time. Fig. 8 represents the single job WAET is decreasing along the increasing of the number of jobs, also an improvement in the performance of ACO over FIFO systems. The job losing rate for both ACO and FIFO is indicated in fig. 9. The dropping curve of ACO represents that there is decrease in the job loss by using ACO algorithm as the number of jobs are increased. The job queue includes jobs with smaller and larger granularity. The simulation result indicates that ACO algorithm is more adaptable for small and large granularity jobs submitted.

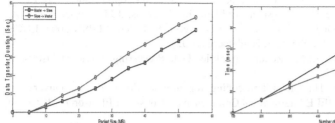

Fig. 6 Network Transfer Time for Large Granularity Jobs

Fig. 7 Varyign Execution Time with Number of Jobs

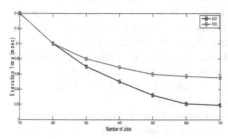

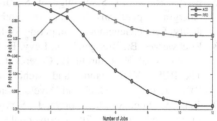

Fig. 8 Single Job WEAT with the Number of Jobs

Fig. 9 Percentage Job Loss as the Number of Jobs Increases

7 Conclusion

In this paper we presented a new paradigm of offering and utilizing the characteristics of cloud computing model and identifying the limitations of Hadoop build in FIFO technique of dealing with the task scheduling for Mobile Cloud Computing model. The proposed Ant Colony Optimization algorithm bridges the gap under Hadoop platform for scheduling the tasks dynamically and enhancing the performance of MCC model for scheduling large number of tasks.

Acknowledgement. I would like to thank the Vice-Chancellor of SASTRA UNIVERSITY for the opportunity and the support provided for this research.

References

1. Klein, A., Mannweiler, C., Schneider, J., Schotten, H.D.: Access Schemes for Mobile Cloud Computing. In: 2010 Eleventh International Conference on Mobile Data Management (MDM 2010), pp. 387–392 (2010)

2. Bianchi, L., Gambardella, L.M., Dorigo, M.: An ant colony optimization approach to the probabilistic traveling salesman problem. In: Guervós, J.J.M., Adamidis, P.A., Beyer, H.-G., Fernández-Villacañas, J.-L., Schwefel, H.-P. (eds.) PPSN 2002. LNCS, vol. 2439, pp. 883–892. Springer, Heidelberg (2002)
3. Dean, J., Ghemawat, S.: MapReduce Simplified Data Processing on Large Clusters. In: OSDI 2004 (2004)
4. Garcia, A., Kalva, H.: Cloud transcoding for mobile video content delivery. In: Proceedings of the IEEE International Conference on Consumer Electronics (ICCE), p. 379 (March 2011)
5. Liu, L., Moulic, R., Shea, D.: Cloud Service Portal for Mobile Device Management. In: Proceedings of IEEE 7th International Conference on e-Business Engineering (ICEBE), p. 474 (January 2011)
6. Dorigo, M., Birattari, M., Stützle, T.: Ant Colony Optimization Artificial Ants as a Computational Intelligence Technique. UniversityLibre de Bruxelles, BELGIUM
7. Buyya, R., Yeo, C.S., Venugopal, S., Broberg, J., Brandic, I.: Cloud computing and emerging it platforms: Vision, hype, and reality for delivering computing, as the 5th utility. Future Generation Computing Systems 25(6), 599–616 (2009)
8. Rochwerger, B., Breitgand, D., Levy, E., Galis, A., Nagin, K., Llorente, L., Montero, R., Wolfsthal, Y., Elmroth, E., Caceres, J., Ben-Yehuda, M., Emmerich, W., Galan, F.: The RESERVOIR Model and Architecture for Open Federated Cloud Computing. IBM Journal of Research and Development 53(4) (2009)
9. Tsai, W., Sun, X., Balasooriya, J.: Service-Oriented Cloud Computing Architecture. In: Proceedings of the 7th International Conference on Information Technology: New Generations (ITNG), pp. 684–689 (July 2010)
10. Vartiainen, E., Mattila, K.V.-V.: User experience of mobile photo sharing in the cloud. In: The Proceedings of the 9th International Conference on MUM 2010 (2010)
11. Yang, H.-C., Dasdan, A., Hsiao, R.-L., Parker, D.S.: MapReduce-merge: simplified relational data processing on large clusters. In: SIGMOD 2007, pp. 1029–1040 (2007)
12. White, T.: Hadoop: The Definitive Guide. O'Reilly, Sebastopol (2009)

Enhancing the Security of Dynamic Source Routing Protocol Using Energy Aware and Distributed Trust Mechanism in MANETs

Deepika Kukreja, Sanjay Kumar Dhurandher, and B.V.R. Reddy

Abstract. A routing protocol for detection of malicious nodes and selection of most reliable, secure, close to shortest and trustworthy path for routing data packets in Mobile Ad Hoc Networks (MANETs) is introduced. Dynamic Source Routing (DSR) protocol [1] is extended and termed as Energy Efficient Secure Dynamic Source Routing (EESDSR). The protocol is based on an efficient, power aware and distributed trust model that enhances the security of Dynamic Source Routing (DSR) protocol. The model identifies the nodes exhibiting malicious behaviors like: gray hole, malicious topology change behavior, dropping data packets and dropping control packets. Monitoring mechanism is suitable for MANETs as it focuses on power saving, has distributed nature and adaptable to dynamic network topology. The new routing protocol is evaluated using Network Simulator 2 (NS2). Through extensive simulations, it has been proved that EESDSR protocol performs better than the standard DSR protocol.

1 Introduction

A Mobile Ad Hoc Network (MANET) is an autonomous group of mobile users that communicate using battery powered devices and is formed when nodes come in close proximity with each other. This implies that there is no need for central administration. Due to the lack of stationary infrastructure, the participating nodes in MANET forward data on behalf of other nodes. Therefore, the functionality of an ad hoc network depends on the forwarding behavior of the participating nodes. Another vital attribute of ad hoc networks is their dynamic topology. The network topology randomly changes

Deepika Kukreja · B.V.R. Reddy
GGSIP University, Delhi, India

Sanjay Kumar Dhurandher
NSIT, Delhi University, Delhi, India

© Springer International Publishing Switzerland 2015
R. Buyya and S.M. Thampi (eds.), *Intelligent Distributed Computing*,
Advances in Intelligent Systems and Computing 321, DOI: 10.1007/978-3-319-11227-5_8

due to node mobility and changes of the surrounding environment. In order to manage with all the above mentioned properties of MANETs, we introduce a distributed trust based model for detection of malicious nodes in the network and selection of a most trustworthy and path between the source and destination. We modified and extended Dynamic Source Routing (DSR) protocol as Energy Efficient Secure Dynamic Source Routing (EESDSR). The proposed secure protocol is suitable for MANETs as it focuses on power saving, has distributed nature and adaptable to dynamic network topology. The trust model is distributed and monitor nodes are selected from time to time so as to adapt with the distinguished features of MANETs. The path selected by EESDSR is not always the shortest but it is next to the shortest path between source and destination since the precedence is given to the most trust worthiest shortest path i.e. a shortest path which is also free from malicious nodes.

The outline of the paper is as follows. In Sect. 2, the summary of the related work on secure routing protocols in MANETs is given. Section 3 provides an overview of important properties of the standard Dynamic Source Routing protocol. Section 4 covers the proposed Energy Efficient Secure Dynamic Source Routing (EESDSR) protocol in detail. In Sect. 5, we present the extensive experimental results and their analysis. Finally, the conclusions are drawn in Sect. 6.

2 Related Work

Dynamic Source Routing was developed and proposed for Mobile Ad Hoc networks by Broch, Johnson and Maltz [1]. There are a number of routing protocols in the literature that were proposed and implemented to secure MANETs [2]. Unlike existing models, EESDSR works on the approach that only few nodes are in promiscuous mode. Marti et al. designed Watchdog and Pathrater method [3] to optimize and improve the technique of packet forwarding in the DSR protocol. It has two major components: Watchdog and Pathrater. Watchdog component detects selfish nodes and Pathrater then uses this information to avoid the detected nodes. CONFIDANT [4] enhances [3] and adds two other components to it: trust manager and reputation system. In [5], Y.Hu et al. proposed ARIADNE, an on-demand secure routing protocol based on DSR for protection against node compromise. It is based upon symmetric cryptography and the distribution of shared secret keys. A solution for black hole attack is proposed by Hongmei Deng [6] where RREQ message is replayed for checking the route from the intermediate node to the destination node. In [7], Pirzada et al. proposed a method for establishing trust based routing in MANETs without requiring a trust infrastructure. Node's trust in [7] is calculated taking in view the packet forwarding behavior. C. Wang et al. [8] proposed a routing algorithm tr-DSR, which is an extension of DSR and is based on nodes' trust and path's trust. The method used in the paper selects the highest trust path used for data transmission. Pirzada et al. [9] modified the DSR protocol such that intermediary nodes act as Trust Gateways that keeps track of trust levels of the nodes in order to detect and avoid malicious nodes. In Pirzada et al. [10] a trust-based model based on direct experience rather than trusted third party is proposed. Sun et al. [11] proposed trust modeling and evaluation methods for secure Ad Hoc routing and malicious node detection. They only considered packet dropping misbehavior to evaluate trust. Huang Chuanhe et al. [12] proposed a trusted routing protocol called Dynamic Mutual Trust based Routing protocol (DMTR), based on DSR protocol that secures the network using the Trust Network Connect (TNC), and improves the path security which is selected by barrel theory. In [13], Azzedine Boukerche et al. proposed a trust based security system called Trust cOmputation and Management System (TOMS). TOMS is installed on every node that monitors its neighboring nodes. The authors proposed a trust-based community model and source selects the route such that each node has the highest trust in the community it belongs. A Guard node based scheme is proposed by Imran Raza and S.A. Hussain [14] to identify malicious nodes in Ad Hoc On-Demand Distance Vector (AODV) protocol. Every node acts as a Guard node for calculating trust of its neighboring nodes and also for determining path trust. Zhongwei Zhang [15] proposed a Fuzzy Logic based Secure Ad Hoc On demand Distance Vector routing protocol (FL-SAODV) to determine security level of a node. P. Narula [16] introduces a method of message security using trust-based multi-path routing. It uses soft encryption techniques and avoids introducing large overheads. The whole message is divided into parts and the parts are self-encrypted. Ayday and Fekri [17] proposed a protocol for availability of data in the presence of insider attacks. The scheme uses game theory to find optimal behaviors of the nodes. X. Li et al. [18] proposed Ad Hoc On-demand Trusted path Distance Vector (AOTDV) multipath routing protocol which gives the shortest trusted path from source to destination and also meets the trust requirement of the data packets. Islam Tharwat et al. [19] proposed a protocol based on agents for selecting the trusted as well as shortest route between source and destination. Jian Wang et al. [20] authors determine the forwarding probability of a node using the concept of attribute similarity. S.K. Dhurandher et al. [21] proposed FACES, in this trust of the nodes is determined by sending challenges and sharing friends' lists.

3 Dynamic Source Routing Protocol

Based on the routing information update mechanism ad hoc routing protocols are classified into three major categories: table-driven, on-demand and hybrid routing protocols. Unlike the table-driven routing protocols in which every node maintains a routing table, on-demand routing protocols execute the path finding process and exchange routing information only when a path is required [22]. Dynamic Source Routing protocol is an on-demand protocol and is composed of two mechanisms that work together to allow the discovery and maintenance of source routes.

3.1　Route Construction

During route construction phase the protocol establishes a route by flooding Route Request (RREQ) packets in the network. The destination on receiving a RREQ packet, responds by sending a Route Reply (RREP) packet back to the source, which carries the route traversed by the RREQ packet received. If an intermediate node receiving a RREQ packet has route to the destination node in its route cache, then it replies to the source node by sending a RREP packet with the entire required route information [22].

3.2　Route Maintenance

When an intermediate node in the path moves away, causing a wireless link to break, a Route Error (RERR) message is generated from the node adjacent to the broken link to inform the source node. The source node reinitiates the route construction procedure.

4　Energy Efficient Secure Dynamic Source Routing Protocol

A Mobile Ad hoc Network is formed by self-governing mobile users carrying energy constrained hand held devices that change the network topology frequently. The energy aware and distributed nature of Energy Efficient Secure Dynamic Source Routing (EESDSR) protocol makes it suitable to employ efficiently in MANET environment as it takes care of the main features of MANETs like dynamic topology, distributed nature and energy constraint devices. EESDSR protocol selects secure and most trustworthy path between the source and destination that is free from malicious nodes. The functioning of the protocol is organized in five phases as follows: Phase I: Selection of monitor nodes Phase II: Trust Model Phase III: Monitoring of nodes' behavior and trust update Phase IV: Trust Propagation Phase V: Route Selection.

4.1　Selection of Monitor Nodes

In the proposed scheme, only a small number of nodes are required to function in promiscuous mode, this results in the cutback of network overhead as compared to the protocols which require all the network nodes to work in promiscuous mode. Nodes working in the promiscuous mode overhear all the transmissions within their range; this requires each node to have high energy capacity. We select Monitor Nodes set (MN set) as in [23] such that all the nodes in the network are either in the MN set or the neighbors of the nodes in the MN set. The rest of the nodes in the network which are not in the MN set are called regular nodes. The monitor nodes themselves are also monitored by the neighboring monitor node/s.The program [23] for the selection of new monitor nodes is executed periodically as well as on demand. The motivation behind the periodic and on demand selection of monitor nodes is the nature of devices or nodes making up the MANET. Monitor nodes are scattered all over the network so that no network node must be left unobserved. As nodes are free to move they change the network topology and require a new set of monitor nodes to be selected from time to time. Monitor nodes work in promiscuous mode; they consume more energy as compared to the energy of regular nodes. In order to circumvent draining down of monitor nodes' energy, a new set of monitor nodes is being selected. The energy of monitor nodes is regularly checked and if remaining energy of any monitor node falls below the energy level required to work in promiscuous mode, the new monitor nodes are selected on demand.

4.2 Trust Model

Each monitor node upholds a trust table that stores trust values of their neighboring nodes. Monitor nodes share the trust values in their trust table with the source, periodically, by sending a packet.

4.3 Monitoring of Nodes' Behavior and Trust Update

During route discovery and data forwarding, MNs evaluate their neighbors based on various parameters. A node can behave maliciously by engaging itself in different malevolent acts. We consider two harmful malicious acts. The first act is that where a node exhibits a malicious behavior by not forwarding or selectively forwarding the packets that it is required to forward to other nodes. In the second malicious act, a node increases the routing overhead by inducing malicious topology change behavior, that is, it first introduces itself in the active route and then leaves the route.

4.3.1 Detection of Packet Dropping and Selfish Behavior

The proposed model differentiates between the selfish behavior and packet dropping misbehavior. A node is said to be selfish if it does not have enough energy and in order to save its remaining energy, it does not forward packets. A node is said to behave malevolently if it has enough remaining energy but willfully drops the packets. Detection of nodes dropping packets is presented in our work [24]. The algorithm presented in [24] updates the trust value of their neighbors depending on the packet forwarding behavior. The detection and trust update function in EESDSR is at variance with that in [24]. In EESDSR, monitoring nodes are selected periodically as well as on demand and nodes inducing packet dropping misbehavior in the network are first detected. The energy of these nodes is then compared with the minimum energy required to forward traffic. If the energy is less than the required energy, then the trust value of these nodes is not decremented as they are being selfish but not malicious. Otherwise, the nodes' trust value is decremented and if the trust value falls below the required threshold, the node is declared as malicious.

4.3.2 Computation for Detection of Malicious Topology Change Behavior

In this malicious act a node forces the network to alter its topology frequently, which eventually causes a high overhead. For the detection of this misbehavior, we necessitate that monitor nodes maintain a table that keeps track of the nodes moving out of the route. Monitor nodes scan this table sporadically, to locate any node that goes away the network frequently. It calculates the mean, μ of the time difference the node leaving the network. Given by:

$$\mu = \frac{(T4 - T1)}{n} \tag{1}$$

Where T4 is the most current time at which a node left the network and T1 is the time at which the node first time left the network after joining and n is the number of times a node left the network till time T4. Based on the value of μ, we calculate a malicious index M(x) that lies between the range [0, 1]. 0 means the highly mobile node inducing malicious topology change behavior and 1 means a low mobility node as requisite for monitor nodes and 0.5 means a node with moderate mobility. The malicious index of node A, M(A) is calculated as:

$$M(A) = \left(1 - e^{\frac{-\mu}{\lambda}}\right) \tag{2}$$

Where λ is the factor by which μ is correlated to the malicious index of node A, M(A). It depends upon the nodes' mobility. In this paper, we consider λ =10 for $\mu = 0$ to 200. The malicious index of node A, M(A) will be near to 0 for a mobile node changing network topology frequently and it is near to 1 if the mobile node is stable. If the computed M(A) of a node comes out to be less than 0.5, then the monitor node decreases the trust of that node by dec_amt.

4.4 Trust Propagation

Monitor nodes work in the promiscuous listening mode, so they learn multiple routes to any destination and store them in their cache. Monitor nodes persistently observe their neighboring nodes and update their trust values according to the behaviors mentioned in Sect. 4.3. Monitor nodes search their cache, periodically for the route from themselves to the source. The route is selected such that there does not exist any malicious node from the monitor node to the source node, using the route selection strategy explained in the next section. The monitor nodes then unicast a packet containing the ID's and trust values of their neighboring nodes through the selected route. Here, the ID and trust values are not appended to any control packet as in [7] and [10], but a new packet is created so that the information about the malicious node is propagated immediately after the detection and new route free from malicious nodes is discovered without delay. The source node on receiving the trust values from monitor nodes, finds the minimum of the previously stored trust value and the recently received trust value for all the nodes. The minimum trust value is then stored in its trust table. The reason for storing the minimum of the two is because of the fact that different monitor nodes may have different trust values for a same node based on their experiences and a node once detected as malicious node should not be given a chance to participate in the route.

4.5 Route Selection

PS-DSR selects the route reactively as PS-DSR is an evolutionary model of standard DSR protocol. The source node makes use of two threshold values for the selection of most trustworthy path between source and destination. The thresholds are termed as Mal_threshold and RTh as clarified below:

Definition 1. Mal_threshold: It is a value at or below which a node is declared as malicious.

Definition 2. RTh: It is a value below which route is not considered as trustworthy.

The source node selects the route for data forwarding as per the following steps:

1. Source node selects the route such that all the nodes in the route have trust value greater than Mal_threshold. Repeat step 1 until a route is selected.
2. Source node computes the route trust of a route as follows:

$$RT_i = \frac{\sum T_i}{n_i^2} \tag{3}$$

Where RT_i is the route trust of route i, $\sum T_i$ is the sum of trust values of all the nodes in route i and n_i is the number of hops in route i.
3. If the computed RT_i in step 2 has value equal or more than RTh, then source node selects the route i. Otherwise go to step 1.

5 Experimental Results and Analysis

5.1 Setup

We used Network Simulator NS-2.34 to evaluate the competence of the proposed protocol. We simulated and match the results of EESDSR against the standard DSR routing protocol. Random waypoint mobility model is used in the simulation setup, in which a node starts at a position with a speed selected between 0-20 m/s without any pause. Packet Delivery Ratio (PDR), packet loss percentage, average end-to-end latency, routing packet overhead and path optimality are used to evaluate the effectiveness of the proposed security scheme. The simulation parameters and trust parameters are listed in tables I and II respectively. CBR rate of 0.2 Mbps and packet size of 1000 bytes ensures congestion free packet transfer and prevents queue overflow. Initial energy 160J for each node is sufficient for providing seamless communication for 140s simulation time. The trust parameters are chosen such that when trust value of a node falls from initial trust to Mal_threshold, this indicates the large number of packet drops that affect the network.

Table 1 Simulation parameters

Parameter	Simulation Value
Simulator	NS-2.34
Examined Protocol	DSR and EESDSR
Simulation time	140 seconds
Simulation area	1500 x 300 m
Number of nodes	60
Transmission range	250 m
Movement model	Random Waypoint
Maximum speed	20 m/s
Pause time	0 seconds
Traffic type	CBR (UDP)
CBR rate	0.2 Mbps
Packet size	1000 bytes
Maximum malicious nodes	25
Initial energy	160 J
rxPower	1 W
txPower	1 W
idlePower	1 W

Table 2 Trust parameters

Parameter	Simulation Value
Trust Range	0.0 to 8.0
Mal_threshold	4.0
Initial trust value	6.0
inc_amt	0.02
dec_amt	0.05
RTh	1.05 and 2.0

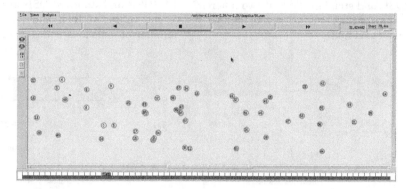

Fig. 1 Network topology of 60 nodes taken for the simulation

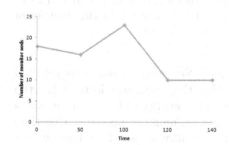

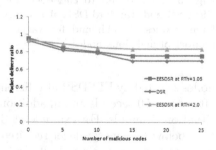

Fig. 2 Number of monitor nodes selected by EESDSR at different times

Fig. 3 Packet Delivery Ratio of EESDSR and standard DSR at two different values of RTh and for varying number of malicious nodes

5.2 Results and Analysis

In this section, we compare the performance results of EESDSR with that of the standard DSR protocol at two different values of route trust threshold (RTh = 1.05 and 2.0). Figure 1 shows a network outlook chosen for the simulation of EESDSR and DSR. Figure 2 shows the number of monitor

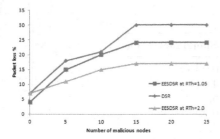

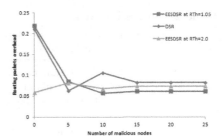

Fig. 4 Packet loss percentage of EES-DSR and standard DSR at two different values of RTh and for varying number of malicious nodes

Fig. 5 Routing packets overhead for EESDSR and standard DSR at two different values of RTh and for varying number of malicious nodes

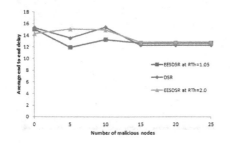

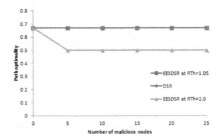

Fig. 6 Average end to end delay for EESDSR and standard DSR at two different values of RTh and for varying number of malicious nodes

Fig. 7 Path optimality for EESDSR and standard DSR at two different values of RTh and for varying number of malicious nodes

nodes selected by EESDSR at different times. New monitor nodes are selected after every 50 seconds and in addition to that they are also selected when the situation demands. For example at time 120 s, energy of few monitor nodes falls below the energy level required to act as monitor node. At this time monitor nodes are selected all over again. Monitor nodes share trust values with the source after every 10 seconds. Figure 3 shows that, the packet delivery ratio using EESDSR is higher than the standard DSR. This is due to the fact that the later does not consider the trustworthy routes. Figure 4, Fig. 5 and Fig. 6 show the Packet loss percentage, routing packet overhead and average end to end delay respectively of EESDSR and standard DSR. EESDSR selects trustworthy and secure route without malicious nodes. This divergence from the routes selected by DSR leads to a lift in the packet overhead as revealed in Fig. 5. The path optimality presented in Fig. 7 for EESDSR at route trust threshold of 1.05 and the standard DSR comes out to be same in our simulation results. The path optimality for EESDSR at route trust threshold of 2.0 is less than that of standard DSR as for higher value of route trust threshold, the protocol becomes more firm and prefers the most trustworthy path over the shortest path.

6 Conclusion

EESDSR guarantee the most reliable, secure and shortest trustworthy path in the sense that the trust value of every node in the selected route is such that all the nodes are benevolent, route include lesser number of hops and the average trust of the route is also greater as compared to that of other routes in the source route cache. Simulations show that in the presence of malicious nodes, EESDSR has higher PDR, lesser packet loss and lesser average end to end latency than the standard DSR. EESDSR is compliant with MANET environment as it is power aware, distributed in nature and adaptable to dynamic network topology. The path selected by EESDSR is next to the shortest path between source and destination since the precedence is given to the most trust worthiest shortest path.

References

1. Johnson, D.B., Maltz, D.A.: Dynamic Source Routing in Ad-Hoc Wireless Networks. In: Imielinski, T., Korth, H. (eds.) Mobile Computing, pp. 153–181. Kluwer (1996)
2. Kukreja, D., Singh, U., Reddy, B.V.R.: A Survey of Trust Based Routing Protocols in MANETs. In: Fourth International Conference on Electronics Computer Technology (ICECT 2012), pp. 537–542. IEEE Press (2012)
3. Marti, S., Giuli, T.J., Lai, K., Baker, M.: Mitigating routing misbehavior in mobile ad hoc networks. In: Proceedings of Sixth Ann. Int'l Conf. Mobile Computing and Networking (MobiCom), pp. 255–265 (2000)
4. Buchegger, S., Boudec, J.: Performance Analysis of the CONFIDANT Protocol: Cooperation of Nodes-Fairness in Distributed Ad Hoc NeTworks. In: Proceedings of IEEE/ACM Workshop Mobile Ad Hoc Networking and Computing (MobiHOC), pp. 226–236 (2002)
5. Hu, Y.C., Perrig, A., Johnson, D.B.: Ariadne: A secure on-demand routing protocol for ad hoc networks. In: Proceedings of the Eighth Annual International Conference on Mobile Computing and Networking (MobiCom), pp. 12–23. ACM Press (2002)
6. Deng, H., Li, W., Agrawal, D.P.: Routing Security in Wireless Ad hoc Networks. IEEE Communications Magazine 40(10), 70–75 (2002)
7. Pirzada, A.A., Datta, A., McDonald, C.: Trust-Based Routing for Ad-Hoc Wireless Networks, pp. 326–330. IEEE (2004)
8. Wang, C., Yang, X., Gao, Y.: A Routing Protocol Based on Trust for MANETs. In: Zhuge, H., Fox, G.C. (eds.) GCC 2005. LNCS, vol. 3795, pp. 959–964. Springer, Heidelberg (2005)
9. Pirzada, A.A., McDonald, C.: Deploying trust gateways to reinforce dynamic source routing. In: Proceedings of the 3rd International IEEE Conference on Industrial Informatics, pp. 779–784. IEEE Press (2005)

10. Pirzada, A.A., McDonald, C.: Trust Establishment In Pure Ad-hoc Networks. Wireless Personal Communications 37, 139–163 (2006)
11. Sun, Y.L., Yu, W., Han, Z., Liu, K.J.R.: Information Theoretic Framework of Trust Modeling and Evaluation for Ad Hoc Networks. IEEE Journal on Selected Areas in Communications 24(2), 305–317 (2006)
12. Chuanhe, H., Yong, C., Wenming, S., Hao, Z.: A Trusted Routing Protocol for Wireless Mobile Ad hoc Networks. In: Conference, I.E.T. (ed.) IET Conference on Wireless, Mobile and Sensor Networks (CCWMSN 2007), pp. 406–409 (2007)
13. Boukerche, A., Ren, Y.: A trust-based security system for ubiquitous and pervasive computing environments. Computer Communications 31(18), 4343–4351 (2008)
14. Raza, I., Hussain, S.A.: Identification of malicious nodes in an AODV pure ad hoc network through guard nodes. Computer Communications 31(9), 1796–1802 (2008)
15. Zhang, Z.: An Intelligent Scheme of Secure Routing for Mobile Ad Hoc Networks. In: 2nd International Conference on Signal Processing and Communication Systems (ICSPCS), pp. 1–6. IEEE (2008)
16. Narula, P., Dhurandher, S.K., Misra, S., Woungang, I.: Security in mobile ad-hoc networks using soft encryption and trust based multipath routing. Sci. Direct Comput. Commun. 31, 760–769 (2008)
17. Ayday, E., Fekri, F.: A protocol for data availability in Mobile Ad-Hoc Networks in the presence of insider attacks. Ad Hoc Networks 8(2), 181–192 (2010)
18. Li, X., Jia, Z., Zhang, P., Zhang, R., Wang, H.: Trust-based on-demand multipath routing in mobile ad hoc networks. IET Information Security 4(4), 212 (2010)
19. Halim, I.T.A., Fahmy, H.M.A., El-Din, A.M.B., El-Shafey, M.H.: Agent-based Trusted On-Demand Routing Protocol for Mobile Ad-hoc Networks. In: Wireless Communications Networking and Mobile Computing (WiCOM) (2010)
20. Wang, J., Liu, Y., Jiao, Y.: Building a trusted route in a mobile ad hoc network considering communication reliability and path length. Journal of Network and Computer Applications 34(4), 1138–1149 (2011)
21. Dhurandher, S.K., Obaidat, M.S., Verma, K., Gupta, P., Dhurandher, P.: FACES: Friend-Based Ad Hoc Routing Using Challenges to Establish Security in MANETs Systems. IEEE Systems Journal 5(2), 176–188 (2011)
22. Siva Ram Murthy, C., Manoj, B.S.: Ad Hoc Wireless Networks: Architectures and Protocols. Prentice Hall (2004)
23. Li, Y., Peng, S., Chu, W.: An Efficient Algorithm for Finding an Almost Connected Dominating Set of Small Size on Wireless Ad Hoc Networks, pp. 199–205. IEEE (2006)
24. Kukreja, D., Miglani, M., Dhurandher, S.K., Reddy, B.V.R.: Security enhancement by detection and penalization of malicious nodes in wireless networks. In: IEEE International Conference on Signal Processing and Integrated Networks, pp. 275–280 (2014)

Dir-DREAM: Geographical Routing Protocol for FSO MANET

Savitri Devi and Anil Sarje

Abstract. Wireless networks form an important part of communication. MANET (Mobile Ad-hoc network) is a much talked about field because of its abundant applications. Majority of MANETs work on RF spectrum as of now. Rise of multimedia applications and smart handheld devices have led to demand for higher bandwidth which has led to research in alternative fields of communication like Free Space Optics (FSO). In this paper we seek to provide a solution for MANET routing by using knowledge of node's location and the directional transmission capability of FSO. This paper aims at developing and simulating a geographical routing protocol that we call Dir-DREAM (Directional Distance Routing Effect Algorithm for Mobility) for a Mobile Ad-hoc network where nodes have multiple FSO Antennas. The proposed protocol uses node's location information and past information of interfaces over which packets from nodes are received for routing. We also perform ns-2 simulations to conduct performance evaluation of Dir-DREAM with FSO interfaces and DREAM over RF interface. Our proposed protocol performs well with multiple FSO interfaces and increases data packet delivery ratio, and decreases end to end delay. The designed protocol is observed to be performing well for varying node speed ranges.

Keywords: Geographical Routing, Free Space Optics, MANET.

1 Introduction

Research in the areas of MANET and FSO (Free Space Optics) [1] are of great importance today for bandwidth intensive applications like multimedia. Today devices have become faster and smarter and demand higher bandwidth. MANETs

Savitri Devi · Anil Sarje
Computer Science & Engineering Department
Indian Institute of Technology, Roorkee
India
e-mail: savitri.negi@gmail.com, sarjefec@iitr.ac.in

© Springer International Publishing Switzerland 2015
R. Buyya and S.M. Thampi (eds.), *Intelligent Distributed Computing*,
Advances in Intelligent Systems and Computing 321, DOI: 10.1007/978-3-319-11227-5_9

have mainly worked on RF spectrum where focus till now has been on omnidirectional forms of communication, in the past some research were done in exploring directional form of RF Antennas for achieving better scalability in networks to support larger number of nodes. The higher demands of growing applications can't be supported by RF with its limited spectrum and bandwidth. FSO has inherent advantages in terms of higher bandwidth and low power consumption.

Directional nature of FSO Antennas leads to advantage of frequency reuse and is less prone to intrusion as any intruder has to be in the LOS (Line Of Sight) to listen to transmission [1]. Current FSO transceivers provide data rates in the ranges of few Gbps for links of up to a kilometer as compared to tens of Mbps and ranges of some tens of meters for RF. Lower power consumption and smaller size make FSO suitable for MANET applications. Along with the stated advantages, FSO also has some shortcomings:

- It requires LOS alignment between the nodes for communication.

- FSO nodes do not perform well during bad weather especially during fog and atmospheric turbulence as it causes optical signal's scattering and absorption leading to high BER (Bit Error Rates) values.

As the directional nature of FSO Antennas make them different from the RF networks so the issues associated with such networks are entirely different too majorly because of directionality and LOS requirement. Some studies of FSO networks suggest that to take maximum benefit of this technology it should be used in a hybrid architecture where both RF and FSO technologies needs to be clubbed together [2]. In such scenarios RF link is suggested to be used to send broadcast data like position information or some control information and as a backup link in case of bad weather conditions when FSO link is unavailable. FSO link is suggested to be used for major data transfers between nodes.

In FSO optical signals are used for information transmission in atmosphere. Basic principle behind FSO is using lasers or multiple LEDs to generate an optical signal that is transmitted through free space which is collected by a photo detector at the receiver side [1]. The signal transmits in free space and provides fiber like data rates without fiber cable laying hassles. Important requirement for this communication to take place is the need for transmitter and receiver to be in LOS. Commercially available point to point FSO links use 850 nm and 1500 nm laser systems. But most of transceivers developed for ad-hoc networks suggest using LEDs because of size and power constraints. Availability of auto alignment circuits and low cost LED array structures make FSO even more promising. Experiments carried out at RPI(Rensselaer Polytechnic Institute) [3],[4] show that such structures perform very well and can overcome the FSO's shortcomings of LOS requirement while using its high bandwidth and long range capability. Prototype of such LED based FSO structures built at RPI shows good capabilities in mobile environment [3].

In this paper we propose a geographical routing protocol Dir-DREAM for FSO Based MANETs. The basic requirement for working of this protocol is that each node should be able to find its location with help of GPS or other localization techniques. The Motivation for this work comes from a study of geographical routing algorithms where it was observed that by having a notion of node's position, routing performance improves by sending data along destination node's direction. The Design of routing protocol for FSO has to take into consideration the directional nature of FSO antennas and the Transmission angle of the antennas where the receiver and transmitter antennas should be in LOS alignment to be able to communicate

2   Related Work

2.1   Geographical Routing Algorithms

The Dir-DREAM belongs to geographical/positional routing group of Routing protocols where node's awareness of its location is a basic assumption. For location information a node uses either additional GPS hardware or localization techniques. Various algorithms in this category are LAR [5], DREAM [6], [7], GPSR [8], EAGRP [9] etc. DREAM [7] uses the concept of Distance between the nodes and node's mobility for dissemination of location packets in network thereby achieving bandwidth efficiency. Later this location information is used to route data using destination node's expected position. GPSR [8] uses greedy and perimeter forwarding to disseminate packets. Whenever a message needs to be sent, the GPSR looks for a node that is closer to the destination than itself and forwards the message to that neighbor node. For topologies with voids it uses concept of perimeter routing. LAR [5] uses expected zone and request zone [5]. Nodes define a request zone for the route request including expected zone, and only nodes lying in that zone forward a route request, this way packets are forwarded. EAGRP [9] uses node distance and node's energy level for choosing next hop. A survey made in [15] about geographical routing protocol states that most of them are greedy methods where some recovery methods like face routing [15] are used when voids are encountered. Other protocols discussed in [15] have other disadvantages, like GOAFR takes assumption of slow nodes while face routing protocols have more overhead. These methods also are not very efficient and have shortcomings. All these geographical routing protocols are for RF networks with omnidirectional antennas while we propose Dir-DREAM for nodes with directional capabilities like FSO in this case. We base our Dir-DREAM protocol on DREAM [7]. There are two main reasons for choosing DREAM. Firstly it takes into account the speed of nodes and their relative impact on information. Second reason is that it works by directionally flooding to last known location which is similar to the need of routing protocol for a FSO based network which also has directional nature.

2.2   FSO Specific Routing Protocols

Some protocols specific to FSO networks are ORRP [10], MORRP [11], and Reconfigurable Routing Protocol [12]. A routing protocol AODVH [2] is there for hybrid RF/FSO network. In AODVH multiple FSO only paths are calculated for data transfer. In this broadcast messages are sent via both RF and FSO networks while data is sent using FSO only paths preferably.

ORRP [10] uses the concept of rendezvous nodes where the routing task is broken in two parts. First part is source to rendezvous node which is reactive and second part is rendezvous node [10] to destination which is proactive in nature. ORRP is designed for static networks. MORRP [11] was designed for mobile networks where rendezvous zones replaced the rendezvous nodes of ORRP as it took into consideration the nodes movement also. MORRP uses a concept of directional routing table where information is stored regarding the nodes that are reachable by a particular interface along with probability of reaching that node via that interface. Another important work in this field is HSR [16] which adds the hierarchical concept by creating clusters physically and defining sub networks logically. Clusters are formed based on their location and cluster heads represents group and manage communication. A hierarchy is formed where cluster heads are iteratively formed. HSR is used in hybrid FSO /RF networks where hybrid nodes generally act as cluster heads and communicate network view amongst them. Our proposed protocol Dir-DREAM is different from these as it is a geographical routing protocol that utilizes node' location information for nodes with FSO interfaces. The protocol proposed here works on nodes having only FSO capability as compared to some of protocols discussed above that work on nodes with both RF and FSO capabilities.

3   Dir-DREAM Description

Our proposed protocol Dir-DREAM is a geographical routing protocol that uses location information for routing. It consists of both proactive and reactive parts where location information dissemination in the network is proactive and sending data is done reactively by using the stored information.

3.1   Node Architecture for FSO Node in ns-2

In ns-2 we created a FSO node by having multiple interfaces on a single node. The transmit receive chain from link layer till wireless-phy layer are replicated. Each network interface is connected to one FSO antenna [13] and the node's link layers are attached to the Dir-DREAM agent. Fig. 1 shows the node structure for such a node with multiple FSO interfaces showing that multiple transmit receive chains exist in a single node which depends on number of FSO interfaces per node. This factor in turn is decided by transmission angle of FSO Antenna. Multiple directional FSO antennas will be used by a node that wants to communicate on longer ranges. Having multiple FSO interfaces instead of a single omnidirectional RF interface will not increase power consumption because of the very low power consumption of FSO devices.

Here Dir-DREAM agent listens on a specified port and sends data to the selected Link layer out of all the Link layers connected to the Dir-DREAM. The important parameters for the FSO Antenna are transmission angle, LOS, antenna position and visibility which we set in our tcl script during simulations. FSO antennas are aligned in the node such that they together cover the area all around. The transmission angle and LOS alignment of the node needs to be set for aligning FSO antennas. The transmission angle of antennas is generally set to 360 / (Number of interfaces).

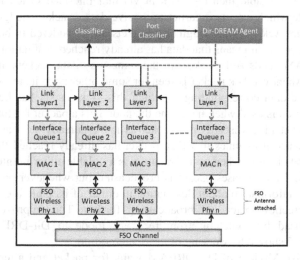

Fig. 1 Modified structure of ns-2 node

3.2 Location Table and Interface Table

In proposed protocol location packets consisting of node's coordinates, timestamp and speed are broadcasted in the network. Location packet act as control packet. This broadcasting is based on DREAM [7] distance and speed effect taking care that location packets are not unnecessarily flooded to nodes that are not affected by other node's movement. Apart from this, Dir-DREAM maintains two tables namely location table and interface table. Location table maintains the location information about nodes and is updated on receiving location packets. In interface table Mac ID of the incoming interface of node is stored and source node's ID which has sent the data or Acknowledgement packet. Other information stored in interface table is hop count and updated time. This information is used to find out a suitable interface to send data to destination or a direction through which a node is reachable. The interface table is updated each time a data or an acknowledgement for the data is received at a node.

3.3   Data Delivery and Acknowledgements

Initially in the network when the Interface table is empty node uses the location information to send data packet. In this case node looks for destination's location in location table. If location information with valid timestamp is available it finds out the angle between the sending node and destination node using location of both nodes. Sending node then finds out its interface that is aligned at that angle and uses that interface to send data. This way data is sent in the approximate direction without flooding the whole network. With time when the interface table gets populated nodes start sending the data using information in interface table thereby decreasing the packet overhead and preventing network flooding. If information is found in interface table then data is sent via that interface which is specified in interface table which was used earlier to receive data or acknowledgment from the specified node. Ack (Acknowledgment) packets are considered to be giving better information because it means that data has already reached destination.

Dir-DREAM sends Ack packets in the same way data packets are sent with a change that whenever Ack packet is seen for some other node it stores the packet id for the source data corresponding to which this Ack packet has been generated. This information is used when next time the node gets a packet for forwarding with the same packet id, it discards the packet as it has already seen an Ack for that packet and hence it assumes that data packet has already reached the destination. Recovery mode is used when no information is available either in interface table or location table. In such cases packet is broadcasted but whenever packet reaches a node that has information about destination in either of its table it can send packet in known direction instead of broadcasting further. This improvement reduces packet overhead by switching from recovery mode to Dir-DREAM mode in between. Fig. 2 shows the working of the Send module.

The Receive Module of Dir-DREAM waits for packet and after checking for duplicate packets it updates Location and Interface tables. Depending on the type of packets received, they are processed and forwarded using the information available in any of the two tables. Fig. 3 shows the working of receive module. If receiving node is the destination node for Ack packets it needs to cancel the resending timers. Location packets are distributed according to the node's specified speed.

4   Performance Evaluation

We simulated Dir-DREAM using ns-2 simulator [14]. We used FSO Propagation Link and FSO Antenna model in ns-2 developed by Mehmat Bilgi and Murat Yuksel at University Of Nevada [13]. First the Dir-DREAM was analyzed to see its performance. We carried out simulations to see how number of FSO antennas and transmission angle affect Dir-DREAM's performance for varying node speeds

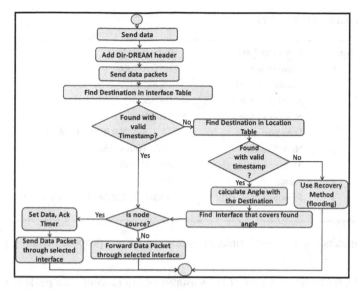

Fig. 2 Send module of Dir-DREAM

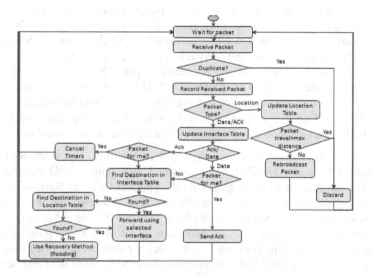

Fig. 3 Receive module of Dir-DREAM

and varying data sizes. Then we compared how Dir-DREAM over FSO links performs in comparison to DREAM protocol working on RF link. We also studied how the number of nodes in the network affects the protocol's performance for varying data size transfers. We used ns-2 for simulations keeping the range of FSO link as default RF range value in ns-2 to ensure that the performance improvements in simulations for Dir-DREAM is not just because of FSO's longer ranges.

Table 1 Simulation Parameters

S.no	Parameter Name	Parameter Value
1	Network Size	1000 X 1000
2	Number of Nodes	16,30,50
3	Simulation Time	100 sec
4	Number of Interfaces	4,6,8,10
5	Data Size	64 Bytes, 256 Bytes, 1024 Bytes
6	Node Speeds	5m/s, 10m/s, 15m/s
7	Transmission Angle	45°, 60°, 90°
8	Range	250 m
9	Propagation Model	FSO (Dir-DREAM), RF(DREAM)

The parameters used for performance evaluation of our proposed routing protocol are:

- Packet Delivery Ratio: It is measured as the ratio of data packets received at destination to the number of data packets sent by the source.
- Routing Overhead: It is measured as routing packets /sec. In Dir-DREAM FSO antennas are used for disseminating Location information. This acts as control packets which need to be broadcast in the network. So we send such packets through all network interfaces on FSO. Routing packet overhead is calculated as packets/sec per FSO interface. If hybrid architecture is used as in AODVH [2] where control packets can be sent through omnidirectional RF antenna this overhead will be reduced.
- Average End to End Delay: It is the sum of all delays for a packet after packet transmission up to its reception at destination measured in seconds.

Results of our simulation are shown in Fig. 4 to Fig. 11. Fig. 4 shows that for simulation of 16 nodes Dir-DREAM performed well with 6 FSO antennas each with transmission angle of 60 degree, with data packet Delivery Ratio of 97.7% at 5m/sec and maintaining it at 83.5% at 15m/sec. Fig. 5 shows that the same FSO configuration also maintained end to end delay between .03 sec at 5m/sec to .16 sec at 15m/sec. This shows that at lesser node density keeping lesser number of FSO antennas perform well. Fig. 6 shows that the packet overhead decreases with larger data sizes because Dir-DREAM uses interface table to forward data and hence for larger data sizes no considerable overheads are used as interface table specify the known interfaces through which destination has been reached in past.

We also compared the Dir-DREAM routing protocol on FSO nodes with DREAM [6] on RF node. Initially we did a simulation with 30 nodes to compare the performance of Dir-DREAM on different FSO node configurations with DREAM [6] on RF node. The results in Fig. 7 shows that for networks with higher density of nodes, having more FSO antennas with lesser Transmission angles is beneficial. Fig. 8 shows that for simulation of 30 nodes our protocol has considerably higher data delivery ratio at speed up to 15 m/sec and protocol overhead doesn't increase much as speed increases.

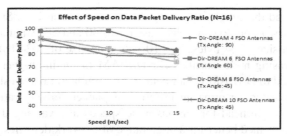

Fig. 4 Effect of speed on data packet delivery ratio

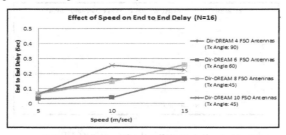

Fig. 5 Effect of speed on end to end delay

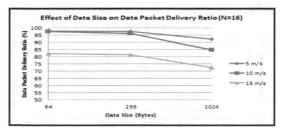

Fig. 6 Effect of data size on data packet delivery ratio

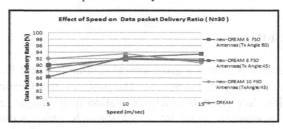

Fig. 7 Effect of speed on data packet delivery ratio

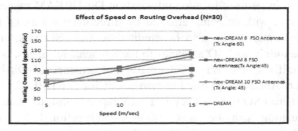

Fig. 8 Effect of speed on routing overhead

We also carried the simulations using different node densities. Starting with 16 numbers of nodes in topology and increased it to 30 and 50. Simulation results in Fig. 9 with data size of 64 bytes, Fig. 10 with data size of 1024 bytes show that even with increasing node densities Dir-DREAM maintains higher data delivery ratio as compared to DREAM for all data sizes. Simulation results shows that Dir-DREAM gives better performance in terms of packet delivery ratio, end to end delay and packet overhead for various node speeds and varying data sizes. The protocol also performed better for a large range of network densities.

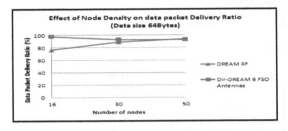

Fig. 9 Effect of node density on data packet delivery ratio (64 Bytes)

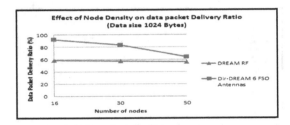

Fig. 10 Effect of node density on data packet delivery ratio (1024 Bytes)

5   Conclusion

In this paper we have proposed a geographical routing protocol Dir-DREAM for MANETs of nodes having FSO Antennas. The protocol uses location information to find out the interface through which destination node can be reached. It also maintains past information about interfaces through which various nodes are reachable. The performance evaluation results show that by using minimum control overhead in terms of location information Dir-DREAM makes good use of FSO's advantages without being effected by directional nature of the link. The increase in performance can also be credited to the FSO link. In the current simulations we didn't set FSO antennas to their reachable ranges, but we believe that by fully utilizing the long FSO ranges, performance of Dir-DREAM will increase further. Another suggested improvement for Dir-DREAM is to use it in hybrid node where FSO link will act as a high capacity link for data transfers and control information can be sent using omnidirectional RF link thereby considerably decreasing packet overhead. Therefore implementing Dir-DREAM over a hybrid RF/FSO network is the future scope of the work as it is the practical approach for using FSO where capabilities of both RF and FSO are utilized. In such network majority of high data traffic will be sent using FSO link and RF will be used for control messages and as a backup link.

References

1. Yan, P., Sluss Jr., J.J., Refai, H.H.: An initial study of mobile ad hoc networks with free space optical capabilities. In: Digital Avionics Systems Conference, DASC 2005, vol. 1. IEEE (2005)
2. Jahir, Y., Atiquzzaman, M., Refai, H., LoPresti, P.G.: AODVH: Adhoc On-demand Distance Vector Routing for Hybrid Nodes. In: IEEE ICC, Cape Town, South Africa (May 2010)
3. Akella, J., Liu, C., Partyka, D., Yuksel, M., Kalyanaraman, S., Dutta, P.: Building Blocks for Mobile Free-Space- Optical Networks. In: Proceedings of IFIP/IEEE WOCN, Dubai, United Arab Emirates, pp. 164–168 (March 2005)
4. Yuksel, M., Akella, J., Kalyanaraman, S., Dutta, P.: Free space optical mobile ad-hoc networks: Auto-configurable building blocks. ACM/Springer Wirless Networks 15(3), 295–312 (2009)
5. Ko, Y.B., Vaidya, N.H.: Location-aided routing (LAR) in mobile ad-hoc networks. Wireless Networks (ACM) 6(4), 307–321 (2000)
6. Camp, T., Boleng, J., Williams, B., Wilcox, L., Navid, W.: Performance comparison of two location based routing protocols for ad hoc networks. In: Proceedings of the IEEE Twenty-First Annual Joint Conference of the IEEE Computer and Communications Societies, vol. 3. IEEE (2002)
7. Basagni, S., Chlamtac, I., Syrotiuk, V.R., Woodward, B.A.: A distance routing effect algorithm for mobility (DREAM). In: 4th Annual ACM/IEEE International Conference on Mobile Computing and Networking, pp. 76–84. ACM (1998)
8. Karp, B., Kung, H.: GPSR: Greedy perimeter stateless routing for wireless networks. In: Proceedings of the 6th Annual ACM/IEEE International Conference on Mobile Computing and Networking (MOBICOM), pp. 243–254 (August 2000)
9. Elrahim, A.G.A.: An Energy Aware WSN Geographic Routing Protocol. Universal Journal of Computer Science and Engineering Technology 1(2), 105–111 (2010)
10. Cheng, B., Yuksel, M., Kalyanaraman, S.: Orthogonal Rendezvous Routing Protocol for Wireless Mesh Networks. In: IEEE International Conference on Network Protocols (ICNP), Santa Barbara, CA, pp. 106–115 (November 2006)
11. Cheng, B., Yuksel, M., Kalyanaraman, S.: Directional Routing for Wireless Mesh Networks: A Performance Evaluation. In: Proceedings of IEEE Workshop on Local and Metropolitan Area Networks (LANMAN), Princeton, NJ (June 2007)
12. Xie, R., Tong, F., Kang, D., Kim, Y.-C.: A reconfigurable routing protocol for free space optical sensor network. In: International Conference on Information Networking (ICOIN), pp. 165–170 (2011)
13. Bilgi, M., Yuksel, M.: Multi-transceiver simulation modules for free-space optical mobile ad hoc networks. SPIE Defense, Security, and Sensing. International Society for Optics and Photonics (2010)
14. The Network Simulator. ns-2, http://www.isi.edu/nsnam/ns

15. Maghsoudlou, A., St-Hilaire, M., Kunz, T.: A Survey on Geographic Routing Protocols for Mobile Ad hoc Networks, Systems and Computer Engineering, Technical Report SCE-11-03-Carleton University (2011)
16. Derenick, J., Thorne, C., Spletze, J.: On the Deployment of a Hybrid Free-space Optic/Radio Frequency (FSO/RF) Mobile Ad-hoc Network. In: IEEE International Conference on Intelligent Robots and Systems (IROS 2005), pp. 3990–3996 (2005)

Extending Lifetime of Wireless Sensor Network Using Cellular Automata

Manisha Sunil Bhende and Sanjeev Wagh

Abstract. The focus of this paper is to use Cellular Automata for simulating a series of topology control algorithms in Wireless Sensor Network using various environments. In this paper, we introduced the use of cellular automata in Wireless sensor networks to control Topology. In the network the sensor nodes are redundantly deployed in the same field. Due to this redundant deployment the many nodes in the network have remained in their active state simultaneously. This causes reduction in the global energy of the network and step-up in the lifetime of the mesh. Thus, the principal aim of the topology control algorithms is to cut the initial topology of wireless sensor network by avoiding noise and hold out the life of the mesh.

Keywords: Cellular Automata, Large scale wireless sensor network, Lifetime, distributed computing.

1 Introduction

Large number of sensors communicate the gathered data wirelessly to a centralized processing station while many sensors are connected to controllers and processing stations directly by using local area networks. Sensor nodes communicate with each other and with a base station (BS) as well. Wireless sensor Networks are collection of sensors collaborating to arrive at particular decision. Information for such decisions is collected from a large number of centrally located nodes known as base stations (BSs). Sensors communicate with each other by using their wireless radios, allowing them to disseminate their sensor data for remote processing, visualization, analysis, and storage systems. This is important because many network applications need hundreds or thousands of sensor nodes. These nodes are often deployed in remote and inaccessible areas. Hence, wireless

Manisha Sunil Bhende · Sanjeev Wagh
University of Pune, India

© Springer International Publishing Switzerland 2015
R. Buyya and S.M. Thampi (eds.), *Intelligent Distributed Computing*,
Advances in Intelligent Systems and Computing 321, DOI: 10.1007/978-3-319-11227-5_10

sensor together with sensing component has additional properties such as on-board processing, communication, and storage capabilities. Due to these enhancements, a sensor node is often responsible for collecting data as well as for in-network analysis, correlation, and fusion of its own sensor data and data from other sensor nodes. When such many sensors monitor large physical environments cooperatively, a wireless sensor network (WSN) is formed. The main component of wireless sensor network is the sensor. The sensors are deployed in the area and these sensors senses information and from the wireless network. According to literature survey, the wireless sensor network is a special kind of network composed of a large number of autonomous sensor nodes geographically scattered on a surface with the ability to monitor the area inside their range and collect data about physical and environmental conditions. Wireless sensor network was first used by the army for tactical surveillance without the need of human presence. There are many applications of the wireless sensor network such as environmental monitoring, industrial process monitoring and control, machine health monitoring etc.

In this paper, we concentrate on Cellular Automata. According to the authors [3] the cellular automata is the idealization of physical systems containing discrete time and space and physical quantities take only a finite set of values. Formally cellular automata are a 4- tuple containing an array of cells or lattice, alphabet, giving the state of cells, neighborhood cells and transition function. The authors also stated the history of cellular automata. A cellular automaton is a discrete computational model which was first used by Jon von Neumann in the 1940s when he was trying to describe self-producing automata. He succeeds by introducing automaton transition rules and starting configurations. After this many researchers used this concept for the further studies. In this paper, we focus on a subset of topology control algorithms and use of cellular automata for simulation.

Most available wireless sensor devices are very constrained in terms of computational power, memory and Communication Capabilities. Wireless sensor networks present a series of serious issues that still need research effort. As sensor nodes are application specific following design objectives are mainly considered while designing a network of sensors.

- **Low node cost** - As nodes are deployed in harsh networks, are many in number and cannot be reused, it is important to reduce the cost of a node. It will lead to reduction of cost of the whole network.
- **Scalability** - As nodes are many in number, the network must be scalable to different sizes.
- **Adaptability**- Some physical changes in the network such as node failure or movement of nodes may affect network topology. So, the network must be adaptive to such changes.
- **Utilization of channel** - Communication protocols designed for sensor networks must use bandwidth efficiently as they have it limited one.
- **Fault tolerance** - Because of harsh deployment environments, sensor nodes are prone to failures. So they must be fault tolerant.

- **Security** - As sensor nodes have properties like storage, computation, network must have some effective security mechanisms to avoid them from unauthorized access or malicious attacks.
- **Self-configurability** - After deployment in the network, sensor nodes must have ability to configure themselves autonomously as well as to re-configure themselves as per topology changes.
- **Connectivity and Coverage-**

The next concept is topology control algorithms. By discovering the minimum configuration of nodes capable of monitoring one for all nodes.

2   Neighborhood Scheme

Different neighborhood selection schemes are used for selecting the neighborhood in cellular automata depending on the performance. Neighborhood selection is made during the update of states over a lattice of cells. Initially the cell itself is considered as its neighborhood, but when the state of a cell is updated by using the transition function the neighborhood is updated. In this paper the effect of different neighborhood selection schemes is studied.

Let us consider different neighborhood selection schemes.

- Moore neighborhood selection

As shown in the figure, the Moore neighborhood selection contains one central cell and eight adjacent cells to it.

- Von Neumann neighborhood selection

As shown in the figure, the Von Neumann neighborhood selection contains one central cell and four adjacent cells to it.

- Margolus neighborhood selection

At each step, the lattice is divided into the blocks of four cells, the cell which belongs to two blocks that alternate during each time step according to whether that step is an odd or even stop.

- Weighted Margolus neighborhood selection

     The Weighted Margolus neighborhood is a variation of the simple Margolus neighborhoods which uses weights. At each time step, each cell decides its state for the next step not only as- cording to the neighborhood block to which it belongs during the current setup, but also according to the neighborhood block in which it belonged during the previous step.

- Slider neighborhood selection

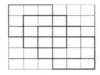

As shown in the figure, the lattice is divided into 3×3 blocks that share one common cell, these blocks alternate every three time step.

3   Topology Control Algorithms

The main purpose of the topology control algorithm is to reduce the global energy wastage. In a wireless sensor network the redundant sensor covers the area due to random deployment. As sensors remain in an active state simultaneously due to the redundancy the global energy of the system is reduced with reduction in the lifetime of the network.

All the topology control algorithms are based on the selection of an appropriate subset of sensor nodes that must remain active. In this paper, we have implemented two topology control algorithms, i.e. Topology Control Algorithms TCA-1 and TCA-2. The main purpose of the topology control algorithm is every alive sensor node, in every step counts its active neighbors. Considering the concept of algorithms, let l be the maximum number of sensor nodes that must remain in an active state during each time step. There are three states of nodes active, idle and dead. Active and idle states imply that sensor node is alive.

The TCA-1 algorithm considers decision active state. The decision regarding to node state is made by node them. The main idea is the selection of an appropriate subset of sensor nodes that must remain active in order to extend the network lifetime, maintaining best possible coverage and connectivity. Let us consider working on this algorithm shortly. Initially, all network nodes are considered as active in assigning them integer value [0, 5] because the nodes are stored in array at each time step. After each time step the integer value decreases by 1. When this value reaches zero, the neighborhood nodes are checked, if at least l number of nodes are found to be active then the node remains deactivated, otherwise it becomes active.

TCA-2 algorithm make decision of the state in terms of predefined categories in which nodes have been classified. TCA-2 is the simulated same as TCA-1 with the extension of the clock. The TCA-2 algorithm contains a clock which is initialized to 2, at each step this clock decreases. In the TCA-2 algorithm classification of the nodes is done according to the criteria that after becoming the timer to zero if there are at least l=2 nodes are active then node is active otherwise it remains idle.

4   Simulation and Results

We have implemented these two algorithms by using different programming environments and studied the results.

System initialization:

1) Deployment of active sensor nodes in the lattice, one sensor per cell. Each sensor node is assigned with 0.8 units of energy, a timer

Randomly initialized with an integer value in [0, 5] and a clock randomly initialized with an integer value in [0, 2].

```
E=0. 8;% initial energy
actv_e=0. 0165;% energy consumption for each action
step
idle_e=0. 00006;%energy consumption for each idle step
%%%% implementation of nodes          '
 n=100;
For i=1: 1:10;
For j=1: 1:10;

%c (I, j) % location of cell or sensor
plot(i,j,'rx','LineWidth',15,'MarkerSize',15);
Hold on;
Draw now;
```

2) At each step, every alive node (*i.e.*, active or idle) works as follows, until it runs out of energy:

A. Decreases its energy, according to rule

B. Checks its clock. If clock value is 0, the node remains in its current state (idle or alive) and sets its clock at position 1 for the next step. If clock value is 1 (or 2), sets its clock at value 2 (or

0, respectively) for the next step and:

a) Decreases its timer by one. If the timer is *not* zero, the node remains in its current state for the current step. If the timer is zero, it is randomly re-initialized in [0, 5] and:

b) Checks the state of the nodes in its neighborhood (including itself).

c) If the sum of the active nodes is more than *l*, the cell Remains/becomes idle during the next step. Otherwise, it remains/becomes

Experimental results are discussed.

Simulations have been developed in Matlab Version 7.0.0.19920 (R14), executed on an Intel Core i3 530 processors at 2.93 GHz with 6144MBytes DDR3 RAM running Windows 7 operating sys- tem. Details on the corresponding implementations of TCA-1 and its variations (assuming Moore, Margolus and Weighted Margolus neighborhoods) can be found in [18].

We have implemented our algorithm on 100 nodes of 10*10 grid.

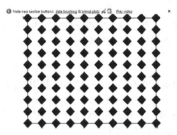

Fig. 1 Grid of 10*10 nodes

1. Results of Moor neighborhood selection

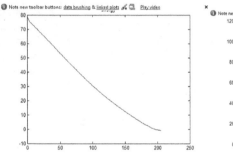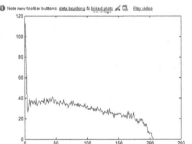

Fig. 2 x axis Network lifetime Vs. Global Energy

Fig. 3 x axis Network lifetime Vs. Network Connectivity

2. Results of Weighted Margolous

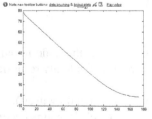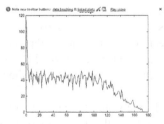

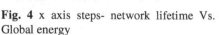

Fig. 4 x axis steps- network lifetime Vs. Global energy

Fig. 5 x axis steps-network lifetime vs. Connectivity

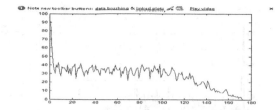

Fig. 6 x axis steps-Network lifetime vs.network Coverage

3. MoorNeighbourhood

Fig. 7 x axis steps-network lifetime vs Coverage

4. Results of TCA1

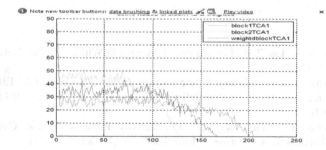

Fig. 8 Comparison of TCA1

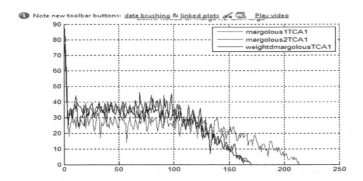

Fig. 9 Comparison of TCA1

5   Conclusion

With the purpose of increasing the network lifetime, this report looks at the problem of determining the utmost number of connected covers in the WSN. Topology control of wireless sensor network using Cellular automata increases the life of the mesh. We have implemented Moor, Von neuman, TCA neighbourhood scemes and compared results of these schemes. Topology management using cellular automata increases the connectivity and coverage of networks based on the selection of randomized or deterministic way.

References

1. Chen, A., Kumar, S., Lai, T.H.: Local Barrier Coverage in Wireless Sensor Networks. Proc. IEEE 9(4) (April 2010)
2. Xing, G., Chang, X., Lu, C., Wang, J., Shen, K., Pless, R., O'Sullivan, J.A.: Efficient Coverage Maintenance Based on Probabilistic Distributed Detection. Proc. IEEE 9(9) (September 2010)
3. Razafindralambo, T., Simplot-Ryl, D.: Connectivity Preservation and Coverage Schemes for Wireless Sensor Networks. Proc. IEEE 56(10) (October 2011)
4. He, S., Chen, J., Li, X., (Sherman) Shen, X., Sun, Y.: Leveraging Prediction to Improve the Coverage of Wireless Sensor Networks. Proc. IEEE 23(4) (April 2012)
5. Liu, C., Cao, G.: Spatial-Temporal Coverage Optimization in Wireless Sensor Networks. Proc. IEEE 10(5) (April 2011)
6. Ammari, H.M., Das, S.K.: Centralized and Clustered k-Coverage Protocols for Wireless Sensor Networks. Proc. IEEE 61(1) (January 2012)
7. Wagh, S., Prasad, R.: Energy Optimization in Wireless Sensor Network through Natural Science Computing: A Survey
8. Nowak, M., May, R.: Evolutionary Games and Spatial Chaos. Nature 359(6398), 826–829 (1992), doi:10.1038/359826a0
9. Applebaum, B., Ishai, Y., Kushilevitz, E.: Cryptography by Cellular Automata or How Fast Complexity Can Emerge in Nature? In: Proceedings of the 1st Symposium on Innovations in Computer Science (ICS 2010), Beijing, January 5-7, pp. 1–19 (2010)

10. Beigy, H., Meybodi, M.R.: A Self-Organizing Channel Assignment Algorithm: A Cellular Learning Automata Approach. In: Liu, J., Cheung, Y.-M., Yin, H. (eds.) IDEAL 2003. LNCS, vol. 2690, pp. 119–126. Springer, Heidelberg (2003)
11. Ganguly, N., Sikdar, B.K., Deutsch, A., Canright, G., Chaudhuri, P.: A Survey on Cellular Automata. Technical Report, Centre for High Performance Computing, Dresden University of Technology, Dresden (2003)
12. Boondirek, A., Triampo, W., Nuttavut, N.: A Review of Cellular Automata Models of Tumor Growth. International Mathematical Forum 5(61), 3023–3029 (2010)
13. Tome, T., De Felicio, J.R.D.: Probabilistic Cellular Automata Describing a Biological Immune System. Physical Review E 53(4), 3976–3981 (1996), doi:10.1103/PhysRevE.53.3976
14. Gardner, M.: The Fantastic Combinations of John Conway's New Solitaire Game 'Life'. Scientific American 223, 120–123 (1970)
15. Chopard, B., Droz, M.: Cellular Automata Modeling of Physical Systems. Cambridge University Press, Cambridge (1998)
16. Ilachinski, A.: Cellular Automata: A Discrete Universe. World Scientific Publishing Co. Pte. Ltd., London (2001)
17. Wolfram, S.: A New Kind of Science. Wolfram Media, Inc., Champaign (2002)
18. Wolfram, S.: Theory and Applications of Cellular Automata. World Scientific, Singapore City (1986)

Computer Network Optimization Using Topology Modification

Archana B. Khedkar and Vinayak L. Patil

Abstract. Computer network optimisation is vital for reducing the costs of the networks, achieving the efficiency, robustness and uniform distribution of the traffic. For network of computers, optimisations are achieved for various aspects such as cost of data transfer, maximum data transfer per unit time, capacity utilisation, uniform traffic distribution. One of the important aspects of computer networks is network topology represented using graph theoretic concepts. Graph theory provides a strong mathematical framework for optimisation of the topology. In this paper, network topology is optimised based on uniform node degree distribution. Uniform degree distribution is achieved just by redistribution of links without any deletion or addition of links to ensure that traffic is uniformly distributed throughout the network and every node is fully utilised without much deviation from average network traffic load. Uniform degree distribution can ensure that traffic cannot be congested, traffic load is distributed and can help in utilisation of the network to its fullest capacity. Owing to these importances, computer network topology is optimised based on Node degree distribution.

Index Terms: Graph theory, Node degree Distribution, Computer Network Optimization, Topology Optimization, Computer Network Traffic, Efficiency, Robustness.

1   Introduction

A building of successful computer network is based on various technical and commercial aspects. Technical aspects focus on speed, data transfer capacity of network at a particular instance of time, individual node and network efficiency in

Archana B. Khedkar · Vinayak L. Patil
Electronics & Telecommunication Department
Trinity College of Engineering and Research, Pune, India
e-mail: {khedkar.archana,patilvl.works}@gmail.com

© Springer International Publishing Switzerland 2015   117
R. Buyya and S.M. Thampi (eds.), *Intelligent Distributed Computing*,
Advances in Intelligent Systems and Computing 321, DOI: 10.1007/978-3-319-11227-5_11

handling data traffic etc. However, aspects and factors such as cost, security, integrity and scalability are to be considered [1]. Most of the network topologies follow the considerations of minimum cost and maximum data traffic with minimum failures (Robustness).Efficiency is the speed in which one can get from one node on the network to another, as well as a suitable and appropriate number of connections for each node. A network that is designed to meet these efficiency factors will generally provide better throughput, is less costly, and is easier to manage. The approaches described in the paper are based on concepts and techniques from computer network analysis, social network analysis [2] and graph theory [3].

Focusing on robustness and efficiency parameters of the network aim is to achieve enhanced network fault tolerance. Although there are various aspects in assessing the robustness and efficiency of a computer network,network architecture, connectivity and topology is considered.

Optimization means finding an alternative to the original system which is to be optimized with most cost effective or highest achievable performance under the given constraints by maximizing desired factors and minimizing undesired ones. That means optimization process improves original system with respect to certain parameters such as time and performance.   Optimization process for computer network refers to best possible balance between network performance and network cost. That is even though performance of computer network is improved its cost should not increase.

In the following sections, the background information and the essential concepts are presented. This includes the introduction of several metrics for measuring the robustness and efficiency of computer networks. These matrices are evaluated and compared using statistical analysis. The results are then provided that were obtained from applying the algorithm to a computer network. The method and its application to a specific example are described in detail and the results are presented.

2   Need of Optimization

Computer network growth result in wide use of internet, where millions bytes of data are exchanges per second. While accessing a web page, from the user computer to the web server a transport layer virtual connection is established. Due to this requested data transfer from the web server to the user computer in the form of packets will starts. These packets follow different switches and routers during transfer process. When many user perform same task, then generated packets will not followed through unique set of switches and routers. Time required by these packets to reach the user computer from the source is referred as the quality of service (QOS). Also data loss that takes place between the source and the user is called as QOS. So optimization of network parameters is done to improve these factors.[4]

For any network there are two possible attacks random attacks and planned attacks. It is important to optimize network that perform better against both of those types of attacks. Because on the average, the optimized robust network has a much lower node- and link-based disconnection ratio, which protect against random attacks. Because of optimization the betweenness values of the nodes and links has been reduced tremendously, hence a targeted attack on such a component will not be nearly as detrimental. On the other hand, as the thinnest nodes on the network have more visibility on the network and they will be better guarded against being taken over by an insider. Hence optimized network provides more robustness and better efficiency, on average, in the case of random failures as well as during targeted attacks.

3   Role of Graph Theory

Graph theory plays an important role in representation of computer networks topology and in determining the various topology parameters in mathematical form. Computer network topology can be optimized based on parameters node degree, shortest paths in network, time, cost and traffic. Out of these parameters node degree parameter is considered. Degree of node in a graph is number of edges incident on vertex. This paper presents structural optimization of computer network.

Initially network topology is represented in the form of set of edges and set of vertices using graph.

A. Mathematical Representation of Graph

Mathematically graphical representation of network topology is given by equation [5]

$$G = (V, E) \tag{1}$$

Set of vertices V and set of edges E are represented as

$$V = \{v_1, v_2, v_3, \dots v_v\} \tag{2}$$

$$E = \{e_1, e_2, e_3, \dots e_e\} \tag{3}$$

B. Properties of Graph

1. Center of Graph

A vertex that has minimum or least distance from all other vertices in the graph is member of vertices those constitute the centre of graph. These types of vertices are

also called central points. It is also a collection of vertices whose longest distance
to all other vertices is the smallest.

$$C(G) = \{u \in V : \in u = rad(G)\} \tag{4}$$

2. Mass of the Graph

If graph $G = (V, E)$, having set of vertices $V = \{v_1, v_2, v_3, \ldots v_v\}$ and set of
edges $E = \{e_1, e_2, e_3, \ldots e_\in\}$, then mass of graph is the mass of links and related
terminal nodes is equal to

$$2 \in = 2|E(G)| = 2E_G \tag{5}$$

3. Volume of Graph

For graph $G = (V, E)$, volume of total number of possible links in the graph G is
equal to

$$|V|(|V| - 1) = v_G(v_G - 1) \tag{6}$$

4. Adjacency Matrix

The Adjacency matrix (A) is formed using direct links from a given node to all
other nodes in the network. For example, if network consists of n vertices, and if
every vertex is directly connected to every other vertices then we write an
adjacency matrix as rows of relation of every vertex to every other vertices as a
matrix equation for A is a generalized adjacency matrix.

$$A = \begin{pmatrix} a_{11} & a_{12} \cdots & a_{1n} \\ a_{21} & a_{22} \cdots & a_{2n} \\ \vdots & \vdots & \vdots \\ a_{n1} & a_{n2} \cdots & a_{nn} \end{pmatrix} \tag{7}$$

C. Undirected Graph

An undirected graph is one in which edges have no orientation. The edge (a, b) is
identical to the edge (b, a), i.e., they are not ordered pairs. An undirected graph
without a self-loop has $n (n - 1)/2$ maximum numbers of edges.

4   Optimization Process

For optimization of computer network, topology is considered. For network
reliability and efficiency node on network must be connected to at least one other
node on network and throughout the steps of optimization this criteria is
considered [6]. Here method is used to configure network topology to obtain
enhanced robustness and efficiency. Steps to be followed during optimization are
given below

   1. Initially network configuration is taken and represented in the form of set of
edges and set of vertices using graph.

2. For this network using the degrees of the nodes, the space of node degrees is divided into three regions. Figure 1 shows the Average Node Degree(AND) distribution along with node degree distribution into three regions.

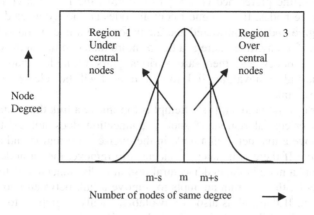

Fig. 1 Distribution of node degree in the network

Where **m** refers to **mean** and **s** refers to **standard deviation**. To define the three regions, the variance and the standard deviation of the network node degrees are calculated and then the region is divided using a given multiple of the standard deviation. The nodes in the region below 1 standard deviation from AND have lower degrees of connectivity and this region is referred as under central region. The nodes in the region within 1 standard deviation of AND have average or close to average degrees of connectivity and this region is referred as central region. The nodes in the region more than 1 standard deviation above AND have higher degrees of connectivity and this region is referred as over central region.

3. For original network metrices which are measures of efficiency and robustness are measured. Also degree values of each node indicating to which of the three regions each node belongs are computed.

4. Link transfer is performed based on certain criteria. As topology changes are implemented two nodes should not be completely disconnected, they should have some sort of connectivity. After link transfer again all metrices and all nodes with their associated degree values are computed [7].

5   Link Redistribution

To obtain better network robustness and efficiency a network is reconfigured, by removing one link from one part of the network and that link is added to a different part of network. The link transfer method uses degrees of nodes in the network, examining the centrality of the network nodes.

A. Link Removal

Excessive connectivity for a node is an indication of unnecessary connections, and therefore unnecessary cost. Excessive and unnecessary connections also cause a network to be overcomplicated. For these reasons, the links between nodes in Region-3 are removed.

In performing the link removal, a link connecting two nodes in the over-central region is removed. In selecting which link among the links in the over-central region to remove, the preference is always to remove the link that connects the two highest degree nodes. If the highest central nodes are not connected, then the next highest degree node is considered and the process continues iteratively until a link is removed. If at any time, there is a tie (or more than one tie) in the degree of the most central nodes and there are various pairs of nodes that could be disconnected via link removal, the link to be removed will be selected at random from among those links.

For the order of link removal, first attempt is to remove a link that connects two nodes in the over-central region. If such a connection does not exist, then it attempts to remove a link between a node in the over-central region and a node in the central region. If there is no such connection to remove, then a node in over-central region and a node in the under-central region is disconnected. If there is no such link available, the program attempts to remove a link between two nodes in the central region. If that link is also not available, finally attempt is to remove a link between a node in the central region and a node in the under-central region. This order is chosen according to the requirement of lowering the degree of higher degree nodes. If there is no link to remove then no link addition is performed, however if a link is removed, then the link addition process begins [8].

B. Link Addition

The nodes with low degree of connectivity are vulnerable as if even one connection from these nodes fail, the node may become disconnected from the network. Furthermore, a very thinly connected node may be an easier target for an insider attacker or for an outsider attacker who might take control of the node. In addition, very thinly connected nodes do not typically contribute much to the connectivity and throughput of a network. For these reasons, links are added to nodes in Region-1.This is another way to improve the network robustness and efficiency by increasing the low centrality of its least central nodes.

In performing the link addition, a link is added to connect two nodes in the under-central region. In selecting which two nodes among the nodes in the under-central region to connect, the preference is always to connect the two nodes with the lowest degree. If the two lowest central nodes are already connected, then the next lowest degree node is considered, and the process continues iteratively until a link is added. If at any time, there is a tie in the degree of the least central nodes and there are various pairs of nodes that could be selected for link addition, the two nodes that are the furthest distance from each other will be linked together [8].

For the order of link addition, first attempt is to add a link that would connect two nodes in the under-central region. If such a connection is not possible, then attempt is to add a link between a node in the under-central region and a node in the central region. If such a connection is not possible, then link is added to connect a node in under-central region and a node in the over-central region. If that is not possible, the program attempts to add a link between two nodes in the central region. If that is not possible, finally attempt is to add a link between a node in the central region and a node in the over-central region. This order is chosen according to the requirement of increasing the degree of lower degree nodes.

6   Robustness and Efficiency Matrices

Metrics are used in assessing and improving network robustness and efficiency are as follows [8],

A. Average Shortest Path Length (ASPL)

ASPL may be used as an indicator of the efficiency of the network and is the average of all of the shortest paths from every node to the other connected nodes. This means that the smaller the ASPL, the more efficient the network is. Given an connected network graph G with the set of vertices V:

   Let $d(v_i, v_j)$ denote the shortest distance between v_i and v_j where $v_i \neq v_j$. Then, given n vertices in G, the ASPL l_G is:

$$l_G = \frac{\sum_{i,j} d(v_i, v_j)}{n(n-1)} = \frac{1}{n(n-1)} \sum_{j=1}^{n} \sum_{k=1, k \neq j}^{n} d_{j,k} \qquad (8)$$

B. Network Diameter (ND)

ND is used as an indicator of the efficiency of a network and is the longest of all of its shortest paths. This can be defined as the 'maximum of all of the minimum paths.

$$ND = Max[Min \ (all \ SPLs) \ ] \qquad (9)$$

   Network's diameter indicate how spread out the network is and the more spread out the network is, the less efficient it is.

C. Average Node Degree Squared (ANDS)

ANDS represents the average of the squared values of the degrees of all of the nodes in the network.

$$ANDS = SSND/N \qquad (10)$$

D. Average Node Betweenness (ANB)

ANB represents the average of the betweenness values for all of the nodes in the network and 'N' is the total number of nodes on the network.

E. Average Link Betweenness (ALB)

ALB is the average value of the betweenness values for all links in the network.

F. Average Node Disconnection ratio (ANDR)

Node disconnection ratio measures the ratio of the network disconnection caused by the removal of a node on the network. ANDR is the average of all of the disconnection ratios caused by removing a node on the network.

G. Average Link Disconnection ratio (ALDR)

Link disconnection ratio measures the ratio of the network disconnection caused by the removal of a link on the network. ALDR is the average of all of the disconnection ratios caused by removing a link on the network.

7   Implementation Flowchart

Its flowchart is explained in fig.3. Aim of topology optimization is to find alternate route in case of any one route fails. That means connectivity between each and every node is achieved directly or indirectly through process of optimization. It makes original network best by distributing node degree uniformly for effective utilization of network. Each node degree value is made near to average value for uniform distribution of traffic by link transfer. It results in uniform distribution of links which makes uniform utilization of network. Uniform traffic distribution throughout the network to its maximum capacity leads to maximum utilization of the network. It also improves robustness and efficiency of the network.

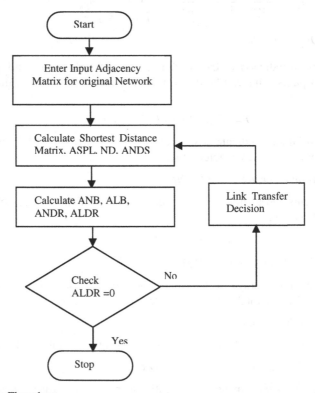

Fig. 2 System Flowchart

8 Results

For illustration purpose network with seven nodes and seven links is considered. This network is represented graphically.

Step 1: Simulation results original network

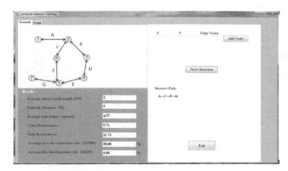

In Simulation window Average Link Disconnection Ratio (ALDR) value is 6.86%. That means if link on the network were failed then 6.86% of the network would be disconnected. Hence link transfer is performed in next iteration.

Fig. 3a Various Metrices calculation

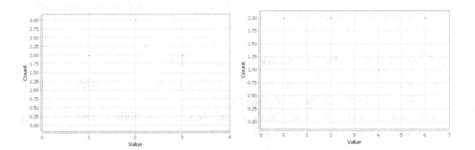

Fig. 3b Degree Distribution   **Fig. 3c** Betweenness Centrality Distribution

Step 2: Simulation results for network after first iteration

- Remove the Link Connecting Node-2 and Node-6(As node 2 and 6 having highest degree of connectivity).
- Link is added between Node-1and Node-6(As node1 and 6 having lowest degree of connectivity).

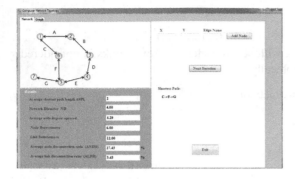

In Simulation window Average Link Disconnection Ratio (ALDR) value is 3.43%. That means if link on the network were failed then 3.43% of the network would be disconnected. Hence link transfer is performed in next iteration.

Fig. 3d Various Metrices calculation

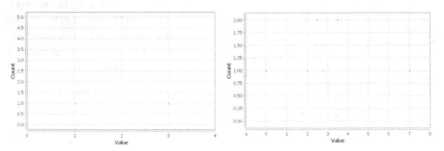

Fig. 3e. Degree Distribution   **Fig. 3f** Betweenness Centrality Distribution

Step 3: Simulation results for network after second iteration

- Remove the Link Connecting Node-5 and Node-6(As node 5 and 6 having highest degree of connectivity).
- Link is added between Node-6 and Node-7(As node 6 and 7 having lowest degree of connectivity).

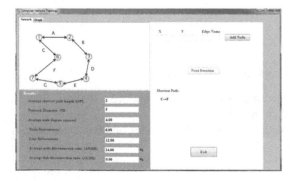

In Simulation window Average Link Disconnection Ratio (ALDR) value is 0%. That means network can not fail or disconnected for the link failure. Hence no need of link transfer and network is said to be optimized.

Fig. 3g Various Metrices calculation

Fig. 3h Degree Distribution   **Fig. 3i** Betweenness Centrality Distribution

9   Conclusion

For topology modification graph structure is used. For this purpose network is initially represented graphically. Then link transfer is done based on node degree parameter. Average node disconnection ratio value is reduced considerably for the network after reconfiguration process. At the end of last reconfiguration there were no more links that met transfer criteria and ALDR is zero. Also degree distribution plot shows degree of nodes will also become near to average value after each iteration. For the last network all nodes degree values are average. This makes uniform distribution of traffic. Therefore network at last instance is more robust and more efficient than initial network, while maintaining all the criteria. Therefore, better and more reliable network configurations are needed to meet the fault tolerance and efficiency requirements of computer networks. Here methodology is described for reconfiguring and expanding computer networks in a manner that will improve the robustness and efficiency of such networks.

References

[1] Rezazad, H.: An Approach to the Development of Intelligent Agents to Assist with Network Configuration Design Problems. UMI, Ann Arbor (2003)
[2] Wasserman, S., Faust, K.: Social Network Analysis—Methods and Applications. Cambridge University Press, Cambridge (1999)
[3] Dekker, A., Colbert, B.: Network robustness and graph topology. In: Proceedings of the 27th Australasian Conference on Computer Science, vol. 26, pp. 359–368 (2004)
[4] Sarwar, S., Mahra, D.: Optimization of Computer Networks. Indian Journal of Computer Science and Engineering (IJCSE) 2(5) (October-November 2011)
[5] Patil, V.L.: Computer Network Optimization. Technical report, Trinity College of Engineering and Research (2012)
[6] Rezazad, H.: A statistical social network approach to computer network optimization. In: Proceedings of the Joint Statistical Meetings. American Statistical Association, Alexandria (2007)
[7] Beygelzimer, A., Grinstein, G., Linsker, R., Rish, I.: Improving network robustness by edge modification. Physica A 357, 593–612 (2005)
[8] Rezazad, H.: Computer network optimization, vol. 3, pp. 34–46. John Wiley & Sons, Inc. (2010)

Mobile Sensor Localization Under Wormhole Attacks: An Analysis

Gaurav Pareek, Ratna Kumari, and Aitha Nagaraju

Abstract. In many application contexts, the nodes in a sensor network may be required to gather information relevant to their locations. This process of location estimation or localization is a critical aspect for all the location related applications of sensor networks. Localization helps nodes find their absolute position coordinates. Like possibly every system, localization systems are prone to attacks. Through this study we intend to do a low-level identification and analysis of broad, large-scale threat to mobile sensor localization systems due to attacks. In this paper, we study the behaviour of some well-known basic localization schemes under by far the most dangerous attacks on localization called wormhole attacks. The network and attacker model assumed in the paper are chosen so that the analysis unleashes the possibility of a resilient solution to the wormhole attacks problem independent of the nodes not under the effect of attacks.

1 Introduction

Localization is a procedure that enables the nodes in a sensor network find their location coordinates in the deployment region. Global Positioning System (GPS) attached to an object also determines the object's location. But the nodes in a sensor network may be deployed in a GPS denied environment or in some application contexts the cost constraints may not allow every sensor node to possess a GPS device. In such cases, the nodes unaware of their location coordinates calculate their locations with the help of a few location aware nodes (also called Anchor Nodes). The Anchor nodes (also called *Beacon* or *Seed* nodes) may be either the nodes having

Gaurav Pareek · Aitha Nagaraju
Central University of Rajasthan Ajmer, India
e-mail: {gauravpareek,nagaraju}@curaj.ac.in

Ratna Kumari
Jawaharlal Nehru Technological University, Hyderabad, India
e-mail: ratna@nigamacollege.com

© Springer International Publishing Switzerland 2015
R. Buyya and S.M. Thampi (eds.), *Intelligent Distributed Computing*,
Advances in Intelligent Systems and Computing 321, DOI: 10.1007/978-3-319-11227-5_12

GPS devices attached to them or the nodes strategically deployed in the region at known locations. The anchor nodes in the network broadcast their location information in the deployment region and all the nodes in the deployment region calculate their location estimates based on the location packets broadcast by the anchor nodes. Localization can be either range-based or range-free depending upon whether the algorithm uses the approximate distance between the normal nodes and GPS enabled anchor nodes or not, respectively. The designer of a localization scheme for mobile sensor networks has to consider the pause time between the two successive moves of a sensor node and the way the direction and distance of the next location relative to the current one is selected. These parameters are determined by the mobility model following which the nodes move in a sensor network. Most popular of these mobility models is Random-Waypoint mobility model (referred to in this paper as RWP mobility model). Most Mobile Sensor localization algorithms use strategies that calculate sample location estimates that upon filtering, approximate to the final location estimates of the sensor nodes.

In section 2 we present the survey of localization algorithms and few classical attacks on localization systems. Section 3 contains the assumptions regarding the network and attacker's capabilities. Section 4 contains the description of the system abstraction and desired analysis results gathered using intensive simulations.

2 Related Work

2.1 Survey of Localization Algorithms

In MCL, the localization problem is modelled as a non-linear stochastic process with a problem to converge the probability distribution of the node being at a particular position given the previous positions $(l_{t-1}^1, l_{t-1}^2, l_{t-1}^3, \dots , l_{t-1}^n)$ of the node and the beacon signals observed in this step $(o_1, o_2, o_3, \dots ,o_n)$. The algorithm consists of three steps namely Initialization, followed by iterative Prediction and Filtering. In initialization step, nothing is known about the position of the nodes (except their maximum velocities)and hence the distribution of nodes is supposed to be completely random in the whole region.Then in the next step prediction, the samples of the nodes positions are calculated depending upon the previous samples, i.e., we calculate $R_{filtered}= \{l_t^i | l_t^i$ where $l_t^i \in R$ and $p(o_t|l_t^i) > 0\}$. Here N is the number of samples (size of the sample set)required for drawing the final position estimate. Now, filtering is done as the last step in which the invalid samples are eliminated from the set of samples depending upon whether the samples receive any one hop or two hop beacon signals or not. That is, now we calculate the complete set of samples by considering the filtered sample set until we get enough samples for that position (set to 50 out of total 80 by Hu and Evans). APIT [14] is another localization scheme suitable more for static wireless sensor networks. It comprises of the following steps:

1. Inside set calculation
2. Centre of Gravity (COG) calculation.

Inside-set calculation step begins with calculating triangles corresponding to the anchor locations (listened to by the node to be localized). If the node to be localized is inside a triangle (the inside test is satisfied), the node adds the triangle to its inside-set. In step 2 of the algorithm, the overlapped area of the triangles is a triangle. The centroid of the triangle so obtained is the approximate location of the node. The algorithm also calculates the departure test to account for mobility of the nodes to be localized. Multilateration Localization Algorithm proposed in [14] is a range-based localization algorithm which calculates the location of a node under consideration by applying statistical methods on the approximate distance (called range) between the node and its anchors and the location of the anchors.

2.2 Attacks on Localization Systems

An attack is called internal if the attacker needs to compromise the key to introduce unauthorised effect to the system [10]. The attack is external if the attacker does not compromise the key to increase the estimation error level. The three types of attacks on localization - Sybil, Replay and Wormhole Attacks are described next.

2.2.1 Sybil Attacks

In this type of internal attack (see figure 1(a)), some compromised nodes acquire legitimate identities and flood their one hop network with some false location information on behalf of compromised identities thereby creating an unauthorized effect.

2.2.2 Replay Attacks

Figure 1(b) shows an external attack (called Replay Attack) in which the malicious node (node 2) replays the packet containing the location information broadcast by a legitimate node (node 1) as it was received at some later time.

2.2.3 Wormhole Attacks

This type of attack involves relay of the beacon messages from one place and replay of the same at some other part of the network giving the sensor nodes in the latter region an illusion that they are in the former region. This type of attack does not need the attacker to compromise the shared secret key. In this, the two wormhole endpoints are deployed in the two distant parts of the network. One to capture the beacon signals broadcast in one region and the other to replay the relayed messages in its region. All the nodes that fall in the communication range of the latter endpoint receive the signal and localize themselves incorrectly.

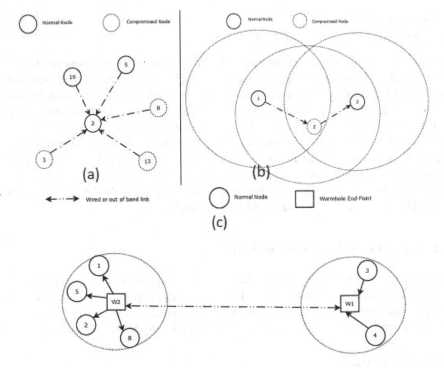

Fig. 1 (a) Sybil Attack: nodes capturing the identities of node 3, 8 and 13 and flooding the node 2 with fake location packets, (b) Replay Attacks: Node 2 replays the location packet received so that node 3 receives the same, (c) Wormhole Attack: W_1 relays the location packets to W_2 which replays them in its region

3 Network and Attacker Model Assumptions

3.1 Network Model

In the network model there are normal nodes and beacon nodes. Both the normal nodes and beacon nodes are mobile. The nodes in the network do not know anything except their maximum velocity. All these nodes are distributed randomly in a 500 × 500 cm^2 region and travel according to RWP mobility model in successive steps.

3.2 Attacker Model

For disturbing the expected accuracy level of the original localization algorithms, we simulate an attacker model which closely imitates the wormhole attacks on localization systems. We assume a scenario in which the two wormhole end-points (A and B) are placed in two distant regions in the network. The end-point A collects all the signals (from both beacon nodes and normal nodes) and relays them to B and then B replays all of them in its region.

4 Simulation Results

In this section, a critical formulation and analysis of the problem of wormhole attacks is presented.

4.1 Increase in Number of Neighbors

Due to the attacker model assumed for experimental verification of the WRMCL algorithm, the number of neighbors that fall in the communication range of a node, increases. The graph in figure 2(a) shows the increase in number of neighbors of every node in the area under attack (on an average) in each step.

4.2 Rise in Estimation Error

Estimation error is the distance between the estimated and actual positions of the nodes in the network. Figure 2(b) 2(c) 2(d) are the comparison of average estimation error in the attacked region in presence of attacks and without attacks for MCL, USC and APIT algorithm.

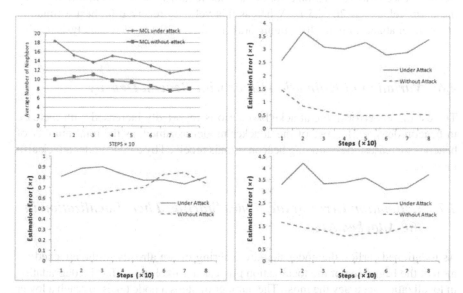

Fig. 2 (a) Increase in average number of neighbors of each node, (b) Rise in Estimation Error with each step for MCL, (c) Rise in Estimation Error with each step for USC, (d) Rise in Estimation Error with each step for APIT. $s_d = 10\%$ $v_{max} = 0.2r$, $r = 50m$

4.3 Rise in Fraction of Nodes Not Having Enough Valid Samples (Only for MCL)

In the absence of enough valid samples, the node will localize itself somewhere near the center of the region. And the algorithm will report: "Not Enough Valid Samples". Figure 3(c) is the comparison of average fraction of nodes in the attacked area having such "inconsistencies "in position estimations in the presence of attacks and without attacks respectively.

4.4 Variation of Fraction of Node Not Able to Fill Their Sample Set with Valid Samples with Beacon Density (Only for MCL)

Since the beacons are also deployed uniformly randomly, the number of colluding beacons will also increase. The variation of the fraction of nodes not able to fill their sample set with valid samples with beacon density is shown in figure 3(d).

4.5 Variation of Estimation Error with Node Velocity

In normal scenario, estimation error of the node localization is at its minimum when the node velocity is 20 (2.5 r). Whereas this velocity received highest estimation error under attacks. Figure 4(d), figure 3(a) and figure 3(b) show the analysis results.

4.6 Variation of Estimation Error with Beacon Density

The estimation error in the attacked scenario is reportedly increased with increase in beacon density. Because of the attacker model assumed, as the total number of beacons increases, the colluding beacon signals received by a node under attack also increases.

4.7 Estimation Error of the Nodes Starting Their Localization in the Attacked Area

As mentioned earlier, the nodes that are entering or are already in the attacked region at the beginning of the localization process are likely to suffer the degradation in localization accuracy the most. The number of steps a node takes to reach a lower stable value of localization error also plays an important role in improving the overall localization accuracy of the deployment. The nodes not only have to get rid of all the colluding links well before the stable step, but also have to regain the localization accuracy using the good links they have.

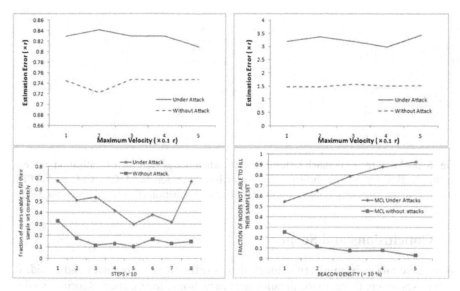

Fig. 3 (a) Variation of Rise in Estimation Error with Maximum Velocity (v_{max}) for (a) APIT, (b) USC ($s_d = 10\%, r = 50m$), (c) Rise in Inconsistencies (Fraction nodes not able to fill their sample set completely with valid samples), (d) Variation of fraction of nodes with inconsistencies with beacon density

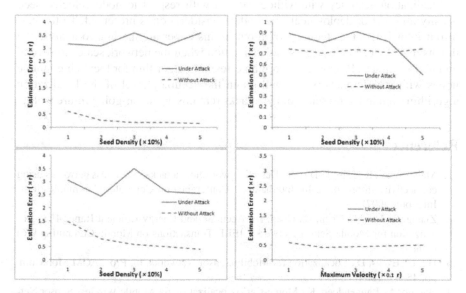

Fig. 4 Variation of Estimation Error with Beacon ratio (Seed Density) for (a) MCL, (b) APIT, (c) USC($v_{max} = 0.2r, r = 50m$), (d) Variation of Rise Estimation Error with Maximum Velocity (v_{max}) for MCL

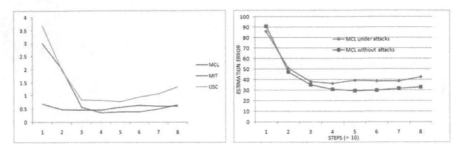

Fig. 5 (a) MCL recovers much faster than USC when nodes start their localization process in the attacked region and further move out of hostile region **for MCL, USC and APIT** ($s_d = 10\%, v_{max} = 0.4r, r = 50m$), (b) Overall Rise in estimation error in each step for MCL

5 Conclusions and Scope

Wormhole attack is an external attack that is difficult to both detect and avoid. It also poses severe threat to localization accuracy. Some localization algorithms with both normal and beacon nodes moving are analysed in under the impact of this attack. We accomplish this by the help of network and attacker model assumed for our study. We come up with the results which can be useful for designing a resilient approach for the localization algorithms analysed in the paper. Analysis of localization accuracy with/without attacks with respect to node velocity, seed density and part of deployment area under consideration is presented. To the best of our knowledge, the studies carried out in this paper are the first to analyse the threats due to wormhole attacks on localization when the network scales to several hundreds of nodes. Proposing a wormhole-resilient algorithm for localizing mobile nodes with mobile anchors that will retain the accuracy level of the localization algorithm even in the presence of the attacks remains as the on-going future work.

References

1. Maheshwari, R., Gao, J., Das, S.: Detecting wormhole attacks in wireless networks using connectivity information. In: International Conference on Computer Communications, Infocom (2007)
2. Zhang, S., Cao, J., Chen, L., Chen, D.: Accurate and Energy-Efficient Range-Free Localization for Mobile Sensor Networks. IEEE Transactions on Mobile Computing 9(6) (June 2010)
3. Hu, L., Evans, D.: Localization for Mobile Sensor Networks. In: Proc. ACM MobiCom, pp. 45–57 (2004)
4. Baggio, A., Langendoen, K.: Monte-Carlo Localization for Mobile Wireless Sensor Networks. In: Cao, J., Stojmenovic, I., Jia, X., Das, S.K. (eds.) MSN 2006. LNCS, vol. 4325, pp. 317–328. Springer, Heidelberg (2006)
5. Rudafshani, M., Datta, S.: Localization in Wireless Sensor Networks. In: Proc. Sixth International Conf. Information Processing in Sensor Networks (IPSN 2007), pp. 51–60 (2007)

6. Lazos, L., Poovendran, R., Capkun, S.: ROPE: Robust Position Estimation in Wireless Sensor Networks. In: The Proceedings of the 4th International Symposium on Information Processing in Sensor Networks, IPSN 2005 (2005)
7. Bettstetter, C., Hartenstein, H., Perez-Costa, X.: Stochastic Properties of the Random Waypoint Mobility Model. ACM/Kluwer Wireless Networks, Special Issue on Modeling & Analysis of Mobile Networks
8. Hu, L., Evans, D.: Using directional antennas to prevent wormhole attacks. In: NDSS 2004, San Diego, CA (February 2004)
9. Zeng, Y., Cao, J., Hong, J., Zhangt, S., Xie, L.: SecMCL: A Secure Monte Carlo Localization Algorithm for Mobile Sensor Networks. In: Mobile Adhoc and Sensor Systems, MASS 2009 (2009)
10. Papadimitratos, P., Poturalski, M., Schaller, P., Lafourcade, P., Basin, D., Capkun, S., Hubaux, J.-P.: Secure Neighborhood Discovery: A Fundamental Element for Mobile Ad Hoc Networking, Security in Mobile AdHoc and Sensor Networks. IEEE Communications Magazine
11. Borg, I., Groenen, P.: Modern Multidimensional Scaling. Series in Statistics. Springer (1997)
12. Sommer, P., Wattenhofer, R.: Gradient clock synchronization in wireless sensor networks. In: Proceedings of the 2009 International Conference on Information Processing in Sensor Networks. IEEE Computer Society (2009)
13. He, T., Huang, C., Blum, B.M., Stankovic, J.A., Abdelzaher, T.: Range-free localization schemes for large scale sensor networks. In: Proceedings of the 9th Annual International Conference on Mobile Computing and Networking, pp. 81–95. ACM (2003)
14. Nagpal, R., Shrobe, H., Bachrach, J.: Organizing a global coordinate system from local information on an ad hoc sensor network. In: Zhao, F., Guibas, L.J. (eds.) IPSN 2003. LNCS, vol. 2634, pp. 333–348. Springer, Heidelberg (2003)

Hash Based Incremental Optimistic Concurrency Control Algorithm in Distributed Databases

Dharavath Ramesh[*], Harshit Gupta, Kuljeet Singh, and Chiranjeev Kumar

Abstract. In this paper, we present a methodology that represents an excellent blossom in the concurrency control environment. It deals with anomalies and assures the reliability of the data before read-write transactions after their successful commitment. This method is based on the calculation of hash value of the data field and compares the current hash value with the previous hash value every time before the write operation takes place. We show that this method overcomes inefficiencies like unnecessary restarts and improves the performance. Finally, this work finds a need for an adaptive optimistic concurrency control method in distributed databases. Thus, a new hash based optimistic concurrency control (*HBOCC*) approach is presented, where it is estimated to produce reliable results. By performing extensive experiments, we epitomize the performance of this method with existing modalities.

Keywords: Concurrency control, Distributed Database, Hash algorithm, Transaction.

1   Introduction

Concurrency control is a problematic instance that coordinates concurrent accesses to a database in a management system. This methodology permits the users

Dharavath Ramesh · Harshit Gupta · Kuljeet Singh · Chiranjeev Kumar
Department of Computer Science and Engineering
Indian School of Mines, Dhanbad - 826004, Jharkhand, India 826004
e-mail: ramesh.d.in@ieee.org,
        {harshit.shift,kooljit.9}@gmail.com,
        k_chiranjeev@yahoo.co.uk

[*] Corresponding author.

R. Buyya and S.M. Thampi (eds.), *Intelligent Distributed Computing*,
Advances in Intelligent Systems and Computing 321, DOI: 10.1007/978-3-319-11227-5_13

to access the database in a multiple programmed manner while preserving the false impression that each user is executing on a dedicated system [19]. Conventionally, the merits of a concurrency control algorithm have been evaluated on the basis of response time and throughput. The main difficulty in achieving this is to prevent the database updates performed by one agent/client from interfering with database retrievals and updates are performed by another agent/client. The problem is deteriorated in a distributed environment because: (i) agents/clients access the data stored in different systems, and (ii) a concurrency mechanism at one system cannot be acknowledged instantly.

In a multi-user environment, a conflict data-access operation may be blocked due to data contention among concurrent operations that requires same data objects that are shared. Consequently, a blocked operation must wait and to be resumed. As distributed database concurrency control is one of the most complex and challenging area of its great significance, where real-time software is necessary for all reliable and critical applications of concurrency control in distributed database. Real-time systems play dominant and critical role, where too many applications involve real-time tasks which often carry remarkable drawbacks in terms of cost and time.

In the recent literature, many significant works have been devoted towards concurrency control methodology where this is required to maintain database reliability from real-time applications used in distributed databases that support (i) description about metadata, (ii) preserving correctness and integrity, and (iii) correct execution of transactions despite failures [20]. Traditionally, databases deal transactions that access the data while preserving its consistency. The goal of transaction processing is to achieve a good throughput along with good response time. On the other hand, real-time database systems can also deal with temporary data. The main difference is that the goal of real-time databases must meet the time constraint mechanism of the transactions. It is important to mention that the real-time instance does not necessarily mean fast [21]. Real-time means the need to manage unconditional time constraints in an expected time period, i.e. to use time methods to deal with constraints associated with tasks and transactions. However, when it comes to ensure the transactions are serializable, the mechanism of conformation is less simple and less straightforward than the blocked mechanism [22]. As we know that the transaction is the basic unit of control which ensures the *ACID* properties of transactions [3]. Every database transaction works with and confirms the *ACID* rules; Atomicity, Consistency, Isolation, and Durability rules. Depending on the transaction types and other factors, different concurrency control mechanisms have been implemented. i) *Pessimistic concurrency control:* If more than one transaction plans to update the same data at the same time, then lock the data and prevents the overlapping update operations of the transactions to access the same data but permit the read operations to access the locked data. ii) *Optimistic concurrency control:* Checks whether the transaction meets the *ACID* rules to the commit state to prevent violation.

This work explores the need for sophisticated optimistic concurrency control methods in real-time distributed database systems. Hence a new optimistic concurrency control approach based hashing algorithm is presented where it is based on reworking to the current workload.

2   Related Literature

There have been many efforts to implement the features usually available in pessimistic concurrency control. Most of these methods are based on Two Phase Locking (2PL). The traditional lock-based pessimistic concurrency control mechanisms like two phase locking protocol of a high priority (2PL-HP) can guarantee the transactions serializability with guaranteed consistency. However, because of restart of transactions, it cannot convince the need of database systems. On the other way, this mechanism believes that the possibility of any two parallel transactions requesting the same database is infrequent [1]. A method related to rollback problem in multi –version concurrency control is proposed to improve the efficiency of the system. But, it has made the system pay the cost of establishing read-write table and scanning it [23].

The real-time database systems require that the number of transactions completed within the deadline for the largest rather than the number of concurrent transaction for execution to maintain the largest [2]. In order to improve system throughput and resource utilization, the processor's idle time should be reduced in the context of reducing transactions waiting time. According to database terminology, a transaction is the basic unit of concurrency control which ensures *ACID* properties of transactions. This is one of the important tasks of the transaction and to check the violation of *ACID* properties caused by the number of concurrent operations [3], [4], [5]. Depending on the amount of conflicts, concurrency control methods are divided into two major categories: pessimistic and optimistic method [3], [4], [5]. Recent studies have focused on processing real-time nested transactions. These studies consider extensions towards locking algorithms [6]. Concurrency control is very important for the consistency and for the correctness of the data under multi-task circumstances. The transaction is the unit of concurrency control and the serialization is the aim of the control. Locking, timestamp, and optimistic controls are all the methods for concurrency control and locking method is the one which is most widely adopted in commercial systems [7], [8]. However, two-phase locking protocol faces severe problems in long duration transactions [9], [10]. A better idea is that the transaction can unlock the data in time when it doesn't need to use it. Yet the idea conflicts with the rule of two-phase locking protocol, measures should be taken to break through the restriction of the protocol to improve the efficiency. However, no kind of method has been proposed to deal with long duration transaction so far [11], [12]. There are many works [13], [14], [15], [16], [17], [18] proposed to solve real-time concurrency control problems in distributed databases. But these works followed their own methodologies and couldn't explore much to solve real –time distributed concurrency control's read-write and rollback instances of the transactions.

In this paper, we highlight an incremental optimistic approach based on hash functionality with MD5 algorithm. By comparing our approach with 2PL – HP [1] and optimistic concurrency control method (OCC) [24] we show the transaction execution is dynamic in nature. The proposed method called hash based optimistic concurrency control (*HBOCC*) algorithm takes less space complexity and responsible for producing correct and reliable results.

3   Projected Methodologies

Optimistic concurrency control methodology is useful for long-running transactions that rarely affect the same occurrence. It exhibits the performance by re-scheduling the data-store elimination on modified reoccurrences. In this study, we use optimistic concurrency control to be applied on real-time distributed database systems with hash based algorithm. Whenever a user requests for the data then the server sends its hash value along with the requested data. If there is more than one transaction, then we initialize the state of each transaction with zero. When a user attempts to write the data, then the computed hash value with the new hash of the data field will be compared. If both the hash values are same then there is no modification of the data during communication. The user can commit the changes and update the transaction state accordingly. At the end, all the transaction states will be compiled for correctness. In case any one of the states is in zero modes then the hash value will be updated and executed again.

3.1   Failure Management

If any transaction fails due to the operation of concurrency control management, instead of aborting the transaction, its client hash value is updated and the transaction is restarted from the previous successful write. For the example, let a transaction 'T' has client hash value 'X' and wants to write a data field, but before its write execution another transaction has written that data. Now it checks the hash value before performing its write operation. If any mismatch occurs, then the transaction will update the client hash value with new hash value of the data and restarts the transaction. We analyze this instance with dependent transaction scenario: If some dependent transactions are considered (i.e. in online reservation scenario – one transaction for ticket booking and one for payment); we divide our transaction into different states. The transaction is successfully committed only if all the states are true. If any of them fails (i.e. payment done but ticket is not booked due to server error), then we redo only the specific state which is failed. If the payment has been done but the ticket has not booked, then instead of restarting the whole transaction the current state of the transaction is saved (i.e. payment done with particular id- '*X*') and whenever the user tries to book again, then the user can use the id to bypass the payment gateway and can book the ticket directly. We analyse this instance by applying our *HBOCC* algorithm.

3.2 Hashing Algorithm

A *hash function* is the one which takes random blocks of data and results a fixed-size bit in the form of a string. The data which is to be encoded are called the *message,* and the hash value is called the *message digest.* The hash function is categorized in to three folds: (i) it is easy to compute for any given message. (ii) It is infeasible to modify a message without changing the hash. (iii) It is infeasible to find different messages with the same hash. Here, we use MD5 hashing, because it is relatively faster with respect to hash functionality. The MD5 algorithm is a widely used hash function producing a 128-bit (16-byte) hash value which is typically expressed in text form as a 32 digit hexadecimal number. MD5 has been utilized in a wide variety of applications and is also commonly used to verify data integrity. The pseudo code of *HBOCC* is depicted in Fig.1. Fig.2 demonstrates about *HBOCC* method using an active diagram as follows: basically, a user receives data and waits until the procedure is completed. Now DBMS can predict whether the requested data arc updated or not. Simply, by comparing the hash id ('*X*') received from the server side, *HBOCC* checks the possibility of matching in the database. Now, MD5 processes a variable-length message into a fixed-length output of 128 bits. It is highly unlikely that, for any two different data fields, the hash value calculated is similar. If the hash values are similar then it is known as Hash collision. The probability of collision can be proved in the following manner.

In probability theory, the birthday problem or birthday paradox [25] concerns the probability can be defined that: in a set of '*n*' randomly chosen people, some pair of them will have the same birthday. By the pigeonhole principle [26], the probability reaches 100% when the number of people reaches 367 (since there are 366 possible birthdays, including February 29). However, 99% probability is reached with just 57 people and 50% probability with 23 people. These conclusions are based on the assumption that each day of the year (except February 29) is equally probable for a birthday. The mathematical terminology behind this problem led to a well-known cryptographic attack called the birthday attack, which uses this probabilistic model to reduce the complexity of cracking a hash function.

The chance of a collision when choosing $n = 2^{32}$ random numbers from a space with $d = 2^{128}$ numbers is approximately. The chance is about 2.7×10^{-20}. This is a very small probability. Hence it is not easy to get hash collisions when using MD5 algorithm. Since here security is not our primary concern and unique identification is of which the probability is very low.

$$p(n;d) \approx 1-\left(\frac{d-1}{d}\right)^{n(n-1)/2}$$

Input: Database with requested data with concurrent transactions.
Output: Ensures the successful modification of required data.

1.   *while* transaction
2.   *if* user Requests Data(user)
      //**Send data and hash value of data to the addressed user**
3.   initialize State Array to 0
      //**Number of writes on different variables is divided into states.**
4.   *for* Each State equals 0
5.   *if* compare (Stored Hash & CurrentHashValue)
6.   *then* do operation
7.   update state to 1
8.   *else*
9.   restore previous consistent data Update Hash;
10.   *if* for each States is Equal To 1
11.   *return* Successful
12.   *else*
13.   *retry*

Fig. 1 Pseudo code of *HBOCC* process

In the case of match, there is no kind of changes made to the requested data. In this way, the server will return success messages to the client and process will be carried forward. In the case of no match, the *HBOCC* will try again and updates the hash value to the client that indicates restart of the transaction for the successful commit. In the case of failure like (shown in section 3.1) payment has been done but ticket has not booked, the value of hash will be newly generated and updated for the same transaction. This is an advantage of *HBOCC*'s approach. We further evaluate and compare its efficiency with *2PL-HP* [1] and *OCC* [24] concurrency control mechanisms in section 4.

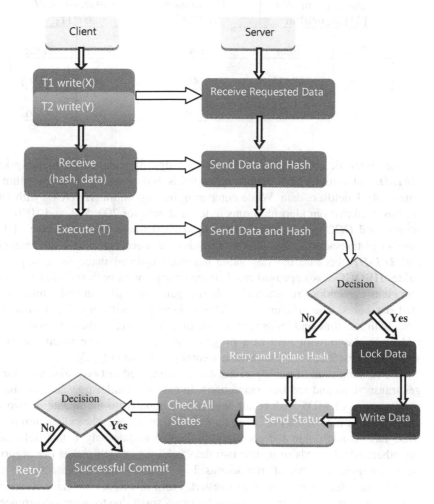

Fig. 2 The sequence of actions of client and server

4   Performance Evaluations

In this section, we compare the performance of our proposed method with the existing approaches. In *OCC* [24] and *2PL –HP* [1], there is a flag field for every data field. Whenever the data field is modified, the flag is incremented by 1 (shown in Table 1). But our algorithm does not use any type of flag for a data field. So, space complexity of our algorithm is almost half of the *OCC* [25] and *2PL –HP* [1], which an important factor in the database.

Table 1 Data values of all algorithms

Database of *OCC* [24] algorithm		Database of *HBOCC*	Database of *2PL-HP* [1]	
Data	Flag	Data	Data	Flag
65	45872	49	86	42361
66	45873	50	87	42362

The optimistic concurrency control (*OCC*) algorithm and 2PL – higher priority (*2PL-HP*) algorithms have to update two fields in database, but our algorithm updates only 1 field i.e. data. While comparing our algorithm (*HBOCC*) with *OCC*, we observed that our algorithm runs little bit slowly for 200, 500, and 1000 transactions and *OCC* occupies less space up to1000 transactions. In case of large number of transactions (from 2000 onwards) our algorithm runs faster than *OCC* and *2PL-HP*, because the flag value has been updated twice for a single data value. HBOCC has improved much in comparing with both traditional methods and gives tremendous results within short expansion of time in real – time distributed databases. The efficiency of *HBOCC* increases with high input of transactions with less time and less space complexity. We validate this setup on a 65 test bed node cluster with Java as a programming language. The result scenario is shown in Table 2 and the comparison scenario is shown in Fig.3.

The running time of our algorithm can be compared in two ways, viz., for larger transactions and smaller transactions. In case of smaller transactions, the calculation time of MD5 hash is not negligible compared to the update of two data fields of older algorithm. For larger number of transactions, the calculation time of MD5 hash is negligible and our algorithm has to update only 1 data field unlike the others which needs to update two data fields i.e. which makes our algorithm faster for larger number of transactions. Fig. 4 shows how a transaction's inter-arrival time affects the transaction's restart. From this figure we can also see that, as interval increases, the rate of restart becomes small due to fewer occurrences of conflicts. But when the interval is not long, the new method is significantly better than traditional methods. The performance of real-time database systems has different targets compared with traditional concurrency control mechanisms. The real-time database systems involve the number of transactions completed within the deadline for the largest. The new method (*HBOCC*) makes the transactions never fail and avoids their unnecessary restart. It saves the system's expense and improves the throughput effectively. As per the data-processing transactions, the new method makes the data elements can be operated on different data items of the transaction without snooping.

Table 2 Transaction execution time for all methods

No. of Trans-actions	Time taken by OCC [25]	Time taken by HBOCC	Time taken by 2PL-HP [1]
10	2ms	2ms	4ms
50	8ms	8ms	9ms
100	16ms	16ms	19ms
200	33ms	35ms	39ms
500	68ms	73ms	82ms
1000	147ms	151ms	154ms
2000	316ms	264ms	318ms
3000	457ms	409ms	479ms
4000	653ms	614ms	682ms
5000	694ms	652ms	712ms

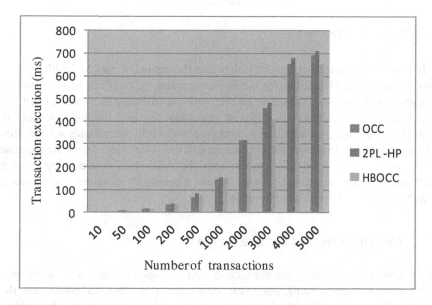

Fig. 3 Transaction execution time Vs number of transactions

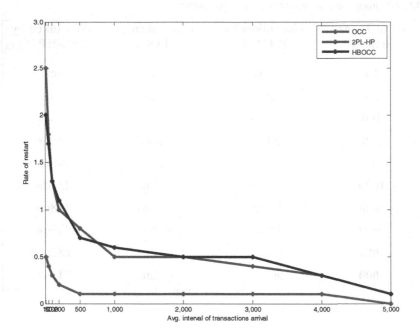

Fig. 4 The restart of Transaction

It collaborates with each other by changing the data elements of the block size. This method of confirmation for the implementation of conflicts to improve the concurrency degree of transaction and the amount of transactions completed before deadline. In summary, in different situations, the new method can flexibly takes the advantage over traditional concurrency control mechanisms with confirmation. It not only improves the concurrency of the system also saves the expense of the system effectively. Compared with the traditional concurrency control mechanisms, the improved Hash Based Optimistic Concurrency Control (*HBOCC*) is quite better on performance.

5   Conclusions and Future Scope

Using hashes for validating the transaction read sets upon commit is a widely used technique for implementing optimistic concurrency control. In conclusion the novelty of the proposed method is shown in taking a hash value of the stored variable and comparing the hash value that is stored at client side with the value of current hash value in a real-time database. Every time, the attribute field updates its hash value which will be changing time by time.

There may be some overheads due to calculation of hash value but it saves much redundant space in the database. At any rate, this methodology can be further extended to represent and can be modify the standard read set validation technique in any fundamental way.

Acknowledgments. This research work is supported by Indian School of Mines (ISM), Govt. of India. The authors would like to express their gratitude and heartiest thanks to the Department of Computer Science & Engineering, Indian School of Mines (ISM), Dhanbad, India for providing outstanding support.

References

1. Garcia-Molina, H., Ullman, J.D., Widom, J.: The Implementation of Database System, pp. 369–377. China Machine Press (2008)
2. Abraham, S., Korth, H.F., Sudarshan, S.: The conception of database. China Machine Press, Beijing (2007)
3. Yang, X.M., Ye, X.-J.: The comparative study on the implementation of concurrency control. Computer Application Research (6), 19–22 (2006)
4. Liu, Y., Wu, H.: Database System Conception. Huazhoang University of Science and Technology Press (1997)
5. Xu: On-line multi-version database concurrency control. Acta Informatica (1992)
6. Guoqiong, L., Yungsheng, L., Lina, W.: Concurrency Control of Real-Time Transactions with Disconnections in Mobile Computing Environment. In: Proceedings of the 2003 International Conference on Computer Networks and Mobile Computing (ICCNMC 2003), pp. 205–212 (2003)
7. Yi, R., Wu, Q.-Y., et al.: A survey of transaction processing technology. Journal of Computer Research and Development 42(10), 1779–1784 (2005)
8. Schuldt, H., Alonso, G., et al.: Atomicity and isolation for transactional processes. ACM Transactions on Database Systems 27(1), 63–116 (2002)
9. Semenov, V., Karaulov, A.: Semantic-based decomposition of long-lived optimistic transactions in advanced collaborative environments. In: Proceedings of the 6th European Conference on Product and Process Modeling, pp. 223–231 (2006)
10. Michael, S., Peter, D., Wolfgang, N.: An environment for flexible advanced compensations of Web service transactions. ACM Transactions on the Web 2(2), 1–35 (2008)
11. Ren, Y., Wu, Q., Jia, Y., Guan, J., Han, W.-H.: Transactional business coordination and failure recovery for web services composition. In: Jin, H., Pan, Y., Xiao, N., Sun, J. (eds.) GCC 2004. LNCS, vol. 3251, pp. 26–33. Springer, Heidelberg (2004)
12. Indrakshi, R., Tai, X.: Analysis of dependencies in advanced transaction models. Distributed and Parallel Databases 20(1), 5–27 (2006)
13. Xia, J.: Distributed real-time database concurrency control system Affairs Strategy Special post-graduate thesis. Journal of SHANXI Normal University (March 2007)
14. Xiong, Y., Bai, S., Li, J.: Distributed Real-Time Database System Management of transaction. Journal of NANCHANG University (June 2008)
15. Zhang, D.: Real-time database research reports to the Alarm and Events a master's degree thesis of North China Electric Power University (April 2008)
16. Liu, H., Zhou, Z., Liao, C.: Historical data processing in real-time database system. Electric Power Automation Equipment, 127–131 (2009)
17. Shanker, U., Misra, M., Sarje, A.K.: A Distributed real time database systems: background and literature review. Distributed and Parallel Databases, 127–149 (2008)

18. Yuan, X., Hua, Z., Fayu, W.: Maintaining temporal consistency in real-time database systems. In: International Conference on Convergence Information Technology (ICCIT 2007), pp. 1627–1633 (2007)
19. Ahn, I.: Database issues in telecommunications network management. ACM SIGMOD Record 23(2), 37–43 (1994)
20. Graham, M.H.: Issues in real-time data management. The Journal of Real-Time Systems 4, 185–202 (1992)
21. Datta, A., Mukherjee, S.: Buffer management in real-time active database systems. In: Real-Time Database Systems Architecture and Techniques, pp. 77–96. Kluwer Academic Publishers (2001)
22. Han, J.-J., Qing-Hua, Essa, A.A.: The scheduling algorithms in transactions of Temporal Constraints in the Real-Time Database. Mini-Micro Systems 26(7), 1229–1232 (2005)
23. Junke, L., Chongqing: The solution to the roll back problem in multi-version concurrency control time stamp protocol. In: Proceedings of 2011 International Conference on Computer Science and Network Technology (ICCSNT), pp. 2803–2806 (December 2011)
24. Rawashdeh, O.A., Muhareb, H.A., Al-Sayid, N.A.: An optimistic approach in distributed database concurrency control. In: Proceedings of 5th International Conference on Computer Science and Information Technology (CSIT), pp. 71–75 (2013)
25. Simion, E., Basista, E., Canal, G., Ziadeh, K.: The Birthday paradox. Operational Research and Optimization (Master EESJI) (December 2012)
26. The Pigeonhole Principle. The Hong Kong University of Science and Technology, Department of Mathematics,
    http://www.math.ust.hk/~mabfchen/Math391I/Pigeonhole

Evaluating Travel Websites Using WebQual: A Group Decision Support Approach

Oshin Anand, Abhineet Mittal, Kanta Moolchandani,
Munezasultana M. Kagzi, and Arpan Kumar Kar

Abstract. The enhanced internet penetration and the increased usage of the facilities by travel industry floods the online arena with websites, and thus generates the urge to find out the best amongst the options and the factors determining it. The article tries to explore the optimally performing travel website in the Indian context based on the evaluation parameters highlighted by WebQual^{TM}. The work would highlight both the leading contributors to reuse of the website as well as their interrelationship. The analysis has been done the Analytic Hierarchy Process for group decision making.

1 Introduction

With the bang of internet all around the world and enhancement in technology, every now and then, both B2B and B2C businesses haven't gone untouched. This is profoundly evident from the inundation of websites catering to online markets, stores, portals etc. Easy entry to this market has led to throbbing competition amongst the players thereby demanding innovation in strategy to attract, indulge and retain the masses. This extends the liability to opt the best amongst the options and optimize them. The scenario is not different in travel industry. Multifarious

Oshin Anand · Abhineet Mittal · Kanta Moolchandani ·
Munezasultana M. Kagzi
Indian Institute of Management - Rohtak, MDU Campus, Rohtak 124001, India
e-mail: oshin.anand@yahoo.co.in, me@abhineet.in,
{moolchandanikanta.moolchandani,Muneza.kagzi}@gmail.com

Arpan Kumar Kar
Department of Management Studies, IV Floor, Vishwakarma Bhavan,
IIT Delhi, Hauz Khas, New Delhi, India, Pin 110016
e-mail: arpan_kar@yahoo.co.in

© Springer International Publishing Switzerland 2015   151
R. Buyya and S.M. Thampi (eds.), *Intelligent Distributed Computing*,
Advances in Intelligent Systems and Computing 321, DOI: 10.1007/978-3-319-11227-5_14

websites are available creating a dilemmatic situation for the users to opt one and even more quandary in being loyal to a single website.

This article focuses on travel websites and investigates website attributes that helps to attract the masses. Websites, being the integral part in the online business, call for continuous innovation and optimization. There have been researches in the past that have found out the factors to be pondered upon when developing a website. The research [1] uses WebQual^{TM} to identify those factors and further find the correlation amongst them. It also analyses the relationship with the frequency of use and reuse of the website. In the following disquisition, we would find out whether the most visited and used websites (top five according to web traffic) depict if they stand appropriate on the factors found using WebQual^{TM}.

A survey has been conducted on the website users, to find the practical importance of these factors with respect to each other amid the customers they are targeting on. A statistical analysis of the responses has been done using Analytical Hierarchy Process (AHP) to conclude the result. The flow of the write-up is as follows- a theoretical background of travel industry, methodologies used; WebQual^{TM} and the investigated factors, AHP, the survey details and result, analysis of the results and finally the conclusion.

2   Literature Review

2.1   Evaluating Website Quality

Website quality measures help to identify and use the factors responsible for the website reuse. For example, WebQual^{TM} was developed based on the Theory of Reasoned Action (Fishbein and Ajzen, 1975; Ajzen and Fishbein, 1980), the Technology Acceptance Model (Davis, 1989), and marketing literatures [1]. Out of an experimental workshop, this instrument was developed to assess the website quality, leading to the generation of quality index (referred to as WebQual^{TM} Index) [11]. Studies concluded WebQual^{TM} instruments for evaluation of websites, on the basis of consumer evaluation to extract several dimensions (broadly four factors, later extended to five factors), keeping the customer in the judge's chair. There are discourse who have tried for quality factors with respect to others like an expert, but since the paramount involvement and effect on the business is due to the customers, the work has finally focused on the parameters pertaining to them. The inception of these factors was **ease of use** and **usefulness,** as they can be concluded out of general perceptions. With further proceedings of digging deep other strongly effecting factors came up like **complementary relationship** and **entertainment. Trust** was a factor later incorporated. The prominent task done in these studies was to further diversify these factors and explore dimensions pertaining to each of them. These factors were traced through dedicated questionnaire for the resolved dimensions [3]. Extended study has further added weightage factor to these dimensions, i.e. a comparative analysis amongst the concluded factors of previous research with the study limited to the US context [1]. Consolidation of these factors along with dimensions is depicted in the following table.

Table 1 Second order factors on WebQual™ used in the study

Initial Higher Level Category	Dimensions	Description
Ease of Use	Ease of Understanding	Easy to read and understand
	Intuitive Operation	Easy to operate and navigate.
Usefulness	Informational Fit-to-task	The information provided meets task needs and improves performance.
	Tailored Communication	Tailored communication between consumers and the firm.
Trust		Secure communication and observance of information privacy.
Entertainment	Visual Appeal.	The aesthetics of a Web site
	Innovativeness	The creativity and uniqueness of site design.
	Emotional Appeal	The emotional effect of using the Web site and intensity of involvement
Response time		Time to get a response after a request or an interaction with a site

From the above stated factors, complementary relationship was dissolved to form two different categories which are trust and response time. Their consideration was due to the result of extended study highlighting factor weightage.

2.2 Travel Industry and Internet Penetration in India

Travel and tourism is an extensive industry worldwide as it appease to both needs as well as likings. Some of recent industry statistics are as follows- The industry has grown with an increase of 4% which is forecasted to appreciate further. With average spending of US$ 1,659 per foreign trip, world outbound travel turnover grew by 6% to US$ 1,571 billion, according to the World Travel Monitor, which covers 90% of world outbound markets. With Asia emerging as a big market (China on top) the internet usage worldwide has also increased to 65%.[2].

According to a report published in Times of India, country's internet population rose to 238.71 million along with the e-commerce market (88%, survey result). The appreciating spread stipulates the increasing importance and usage of internet for need fulfillment purposes which incorporates shopping, social media, survey and search (reviews) which further indicated trust on online data. This even extends to travel website.

3   Focus of Study

There has been studies concentrating on exploring the WebQual^{TM} parameter and finding out their prioritization sequence in a specific context (US) along with their interrelationship. This study aims at using these parameters to judge upon travel website options besides finding out whether the parameters respond in a similar manner in Indian context.

The study shortlists websites for further discourse on the basis of web traffic, the details of which is depicted below:

Travel Websites (picked up)

IRCTC: Indian Railway Catering and Tourism Corporation, abbreviated to IRCTC, is a subsidiary of the Indian Railways that handles the catering, tourism and online ticketing operations of the railways. Within a short span of its going online, the IRCTC website had become the largest and the fastest growing e-commerce website in the Asia-Pacific region, with about six lakh registered users as of 2013. (Wikipedia).

Make my trip: It is an online travel company headquartered in Gurgaon. The company provides online travel services including flight tickets, domestic and international holiday packages, hotel reservations and rail and bus tickets. The website is showered with award for best travel portal, 2012 by Outlook Traveler's Reader's Choice Award (website).

Goibibo: Goibibo is one of the leading online travel sites in India that enables travelers to buy air and bus tickets, hotels and holidays. It is a part of the MIH India Group and was launched in September 2009 (official website).

Yatra.com: Yatra.com is an Indian online travel agency and a travel search engine based in Gurgaon, Haryana, founded by Dhruv Shringi, Manish Amin and Sabina Chopra in August 2006. In April 2012, it was the second largest online travel website in India, with 30 per cent share of the ₹370 billion (US$5.9 billion) market for all online travel-related transactions (Wikipedia).

Cleartrip.com: Cleartrip.com is one of the leading domestic travel portal for cheap flights, airline tickets, hotel bookings, travel guide, travel information (Wikipedia).

These websites have been selected according to web traffic recorded for each of them.

4   Research Methodology

The objective of the study is to arrive at conclusion about the best travel website, specifically in Indian context and therefore the competing websites are those which has highest web traffic in India. The data collection (Primary) is done through a survey conducted online using a structured questionnaire. Substantial additions were done through secondary data achieved through various websites and literature. The sample is around hundred in number with the sample unit comprising of people who visited these website and responded to the questionnaire. The technique incorporated was systematic cluster sampling.

The study incorporates two different surveys, one which measures relative importance of one parameter over another considering all possible combinations of five factors, and second measures the performance of each of the selected websites on all the twelve dimensions which finally club to form the five major factors.

The computational part involves the usage of AHP for the prioritization purpose where the obtained weights and the responses are used to arrive at the final decision of the study. The AHP, developed and introduced by Saaty [4, 6], is a decision-making approach incorporating varied criteria. The technique helps in prioritizing amongst varied parameters considered for evaluation, which can be effectively made out from different studies [2, 9]. To capture qualitative as well as quantitative attributes, fundamental scale has been proven to be good enough by previous studies. Usage of fundamental scale of absolute numbers by AHP makes it a robust process which can be validated from physical as well as decision problem experiments [4, 6]. It converts individual preferences into ratio scale weights that can be combined into a linear additive weight for each alternative. It is structured technique, used for complex decision analysis and is extensively used which is evident from the following studies [7, 8, 9, 10]. AHP computational flow can be depicted diagrammatically as follows:

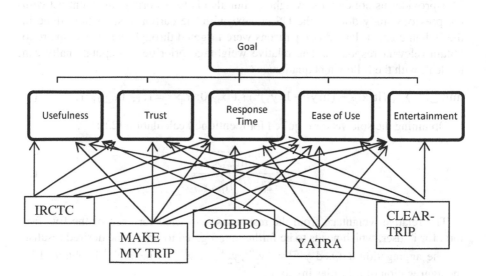

Fig. 1 Hierarchical structure of preference elicitation using AHP

The data source for the study was survey which is done through a structured questionnaire. The sampling was systematic cluster sampling with sampling unit as an individual who have used these websites. The analysis of data and the subsequent outcome is highlighted in the following sections of the paper.

5 Computation Methodology

To evaluate the final decision, the five WebQual[TM] factors (usefulness, ease of use, entertainment, trust and response time) are utilized to arrive at the conclusion. Since AHP is conclusive using associated weightages with different parameters in the study, the survey records responses accordingly.

The responses recorded were of two types

a) Responses recording relative importance of one WebQual[TM] parameter to that of others which provided respective weightages associated with each of the five parameters.

b) The second response weighs the performance of the selected websites, on several dimensions pertaining to the five factors.

5.1 Calculation of Relative Weightage

The questionnaire designed in the survey, records relative response, which is used to calculate the weightage or importance of one parameter with respect to others. This provides us not only the weightage but also helps us bring out a contrast with the previous study done in the US context with the current disquisition, done in the Indian context. Paired comparisons were targeted through the questionnaire to obtain relevant responses. The relative weightage priorities computationally can be dealt with the following equations.

$$\min \sum_{i=1}^{k} \sum_{j>i}^{k} (\ln x_{i,j} - (\ln y_i - \ln y_j)^2) \text{ s.t. } x_{ij} \geq 0; \ x_{ij} \times x_{ji} = 1; \ _i \geq 0, \sum y_i = 1. \tag{1}$$

Obtaining the sole vector required for mentioned calculation [22]:

$$y_i = \frac{\sqrt[1/k]{\prod_{j=1}^{k} x_{i,j}}}{\sum_{i=1}^{n} \sqrt[1/k]{\prod_{j=1}^{k} x_{i,j}}} \tag{2}$$

The relative weightage from the responses, extracted for each of the five factors, for i[th] user, which needs to be further aggregated to obtain the desired results.

The aggregation defined $y^{(a)} = (y_1^{(a)}, y_2^{(a)}, \ldots y_r^{(a)})$ where $y_i^{(c)}$ is obtained by the aggregation of priorities involving;

$$y_i^{(a)} = \frac{\prod_1^n (y^{(k)}{}_i)^{\Phi_i}}{\sum_1^r \prod_1^n (p^{(k)}{}_i)^{\Phi_i}} \tag{3}$$

5.2 Calculation of Individual Preference in Dimensions

Hundred responses recorded, are clubbed according to their dimensions pertaining to each of those factors, which is further summed up to conclude the response with regards to a specific website in the study.

5.3 Calculation of Final Result

These weightage are used to arrive at the conclusion using AHP calculation. The technique uses pairwise comparison which is incorporated in the first calculation. According to the nomenclature this hierarchy starts with the objective, which leads to the criteria which are the five factors. These five factors have two aspects- there relative importance amongst the respondents and their individual scores with respect to websites. Finally these criteria lead to the prioritization of the alternatives. The matrix, computed from the analysis of two responses, is used to compute the weightages associated with each website and finally the weightages are used to finally decide the best website.

6 Result and Discussion

The first prominent amalgamation of the calculations is as follows:

Table 2 Relative weightage of each factor measuring quantifying importance

Usefulness	Trust	Response Time	Ease of Use	Entertainment
0.062	0.379	0.257	0.172	0.129
Rank 5	Rank 1	Rank 2	Rank 3	Rank 4

The above table represents the relative weightage of each factor corresponding to the other. The research [1] concludes usefulness, response time and entertainment of maximum importance and trust does not score anywhere. Diverging from the previous result, in this study, trust scores high with respect to others and usefulness the lowest. This brings out a diversifying contrast due to the context of the studies conducted.

Here is the consolidated outcome of the performance of each website under study against each factors building the criteria for the decision:

Table 3 Performance of each website with respect to the factors

	Useful-ness	Trust	Response Time	Ease of Use	Enter-tainment
IRCTC	0.202	0.231	0.118	0.202	0.182
Make My Trip	0.220	0.207	0.233	0.214	0.222
Goibibo	0.193	0.187	0.220	0.197	0.199
Yatra	0.198	0.193	0.216	0.197	0.204
Clear-Trip	0.188	0.183	0.214	0.190	0.193

The above table consolidates the calculation of weightages of each individual factor with respect to that of each website. These factors were responded separately on the different dimensions mentioned above, which again was analyzed. The above table actually finds the contribution of each website to each factor. Since identical sample is responding to both the questionnaire, it can be said that these are the contributions to the weightages associated with each factor. Therefore, the consolidation of the above two tables result into the following

Table 4 Final weighted scores of websites; utilized to rank

S.N.	Website	Final score	Rank
1	IRCTC	0.189	5
2	Make My Trip	0.217	1
3	Goibibo	0.199	3
4	Yatra	0.201	2
5	ClearTrip	0.194	4

It shows that the website scoring highest according the weightages assigned is Make My Trip which does not align to that of the conclusions through web traffic that ranks IRCTC highest which is contrastingly ranked lowest in the analysis.

7   Conclusion

The survey for the work is done in India, which has a different cultural context which is profoundly depicted in the result that draws opposites to that of the previous work done in US. Trust factor gains importance (table 2). When consolidating all the factors IRCTC of highest web traffic falls apart in scoring on response time thereby dropping down in the overall summary (table 4). Usefulness still remains a major contributor, (highest make my trip) providing for its increased weightage. Factors apart from trust are scored fairly well for Make my trip website, making it optimal.

It is important to note that trust and response time are the two most important evaluation criteria, from among the different WebQual dimensions, and have been rated as being most critical, by the participants of the survey. Indeed this is an interesting insight and throws a lot of light into the contextually of usage. This context, that of travel, being a leisure activity, and easily addressable by alternative intermediaries, trust on the source and the time taken for the delivery of actual outcome plays an extremely important role in the perceived quality of the outcome, that of travel planning.

Further work can be done to optimize the result further using improved AHP techniques like incorporation of fuzzy concept. Besides, the study is capped by a specific context (India) and some of the other popular travel websites like Expedia not having significant web traffic in the country, are excluded. Only one parameter was considered for the selection of websites, which could be changed to multiple parameters to get better results.

References

1. Loiacono, T.E., Chen, Q.D., Dale, G.L.: WebQual™ Revisited: Predicting The Intent To Reuse A Web Site. In: Eighth Americas Conference on Information Systems (2002)
2. World travel trends report: ITB (December 2013)
3. Loiacono, T.E., Chen, Q.D., Dale, G.L.: WebQual™ Revisited: Predicting The Intent To Reuse A Web Site. In: Eighth Americas Conference on Information Systems (2002)
4. Saaty, T.L.: A Scaling Method for Priorities in Hierarchical Structures. Journal of Mathematical Psychology 15, 57–68 (1977)
5. Saaty, T.L.: The Analytic Hierarchy Process. McGraw-Hill International (1980)
6. Saaty, T.L.: Fundamentals of Decision Making and Priority Theory with the AHP. RWS Publications (1994)
7. Golden, B., Wasil, E., et al.: The Analytic Hierarchy Process. Applications and Studies (1989)
8. Kumar, S., Vaidya, O.S.: Analytic Hierarchy Process-An overview of applications. European Journal of Operational Research 169(1), 1–129 (2006)
9. Liberatore, M., Nydick, R.: The analytic hierarchy process in medical and health care decision making: A literature review. European Journal of Operational Research 189(1), 194–207 (2008)
10. Omkarprasad, V., Sushil, K.: Analytic hierarchy process: an overview of applications. European Journal of Operational Research 169(1), 1–29 (2006)
11. Stuart, B., Richard, V.: WebQual: An Exploration of Web-site Quality School of Management. University of Bath, Bath BA2 7AY, United Kingdom
12. Evangelos, T., Stuart, M.H.: Using the analytic hierarchy process for decision making In engineering applications: some challenges. Inter'l Journal of Industrial Engineering: Applications and Practice 2(1), 35–44 (1995)

13. Escobar, M.T., Moreno-Jiménez, J.M.: A note on AHP group consistency for the row geometric mean priorization procedure Universidadde Zaragoza, Gran Via 2, Zaragoza 50005, Spain
14. Bauer, C., Scharl, A.: A Classification Framework and Assessment Model for Automated Web-site Evaluation. In: Proceedings of the Seventh European Conference on Information Systems, pp. 758–765 (1999)
15. Wikipedia, http://en.wikipedia.org/wiki/Category:Online_travel_agencies
16. Travel Wire Asia, http://www.travelwireasia.com/2011/06/top-10-travel-websites-in-india/
17. Shim, J.: Bibliography research on the analytic hierarchy process (AHP). Socio-Economic Planning Sciences 23(3), 161–167 (1989)
18. Sureshchander, G.S., Rajendran, C., Anatharaman, R.N.: The relationship between service quality and customer satisfaction: a factor specific approach. Journal of Services Marketing 16(4), 363–379 (2002)
19. Zahedi, F.: A Simulation Study of Estimation Methods in the Analytic Hierarchy Process. Socio-Economic Planning Sciences 20, 347–354 (1986)
20. Abels, E., White, M., Hahn, K.: Identifying user-based criteria for Web pages. Internet Research: Electronic Networking Applications and Policy 7(4), 252–262 (1997)
21. Ngai, E.W.T.: Selection of website for online advertising using AHP. Information & Management, 1–10 (2002)
22. Crawford, G., Williams, C.: A note on the analysis of subjective judgement matrices. Journal of Mathematical Psychology 29, 387–405 (1985)

Location-Based Mutual and Mobile Information Navigation System: Lemmings

Simon Fong, Renfei Luo, Suash Deb, and Sabu M. Thampi

Abstract. In this paper, the design of an asynchronized messaging system where a user may proactively seek for answers or advice by depositing a question on the messaging system is presented. The messaging system will automatically disseminate the question which is related to a specific location, to a group of users who are either within the proximity currently or have just recently been there. The users are supposed to help each other; karma or some sort of point/score system can be added to distinguish the most helpful users in the community. A name is suggested for this system called "Lemmings" which stands for Location-based Mutual and Mobile Information Navigation System; it is based on a classical video game where a group of creatures have to work and win through the puzzle game together. The design presented is modular and it comprised of the logics, the data flows, and the database schema.

Keywords: Location-based service, collaborative system.

1 Introduction

Nowadays, several location-based services (LBS) allow their users to take advantage of information from the Web about points of interest (POIs) such as

Simon Fong · Renfei Luo
University of Macau, Taipa, Macau SAR
e-mail: ccfong@umac.mo

Suash Deb
Cambridge Institute of Technology, Ranchi, India
e-mail: suashdeb@gmail.com

Sabu M. Thampi
Indian Institute of Information Technology & Management, Kerala, India
e-mail: sabu.thampi@iiitmk.ac.in

© Springer International Publishing Switzerland 2015   161
R. Buyya and S.M. Thampi (eds.), *Intelligent Distributed Computing*,
Advances in Intelligent Systems and Computing 321, DOI: 10.1007/978-3-319-11227-5_15

cultural events or restaurants. The new generation of Mobile Recommender should be collaborative, provide information taking into account user preferences, provide peer-advices and peer-rapports, in addition to location information, that contribute to define the context of use, for example, time, day of week, weather, user activity or means of transport.

The new research aims at developing mobile social advisor systems able to identify user preferences and information needs, thus suggesting information in two types of modes: Passive-mode – personalized recommendations related to POIs in the surroundings of the user current location, like a Mobile Yellow Pages. The proposed systems achieve the following goals: 1. identify user preferences and needs to be used in the information filtering; 2. to exploit the ever-growing amount of information from social networking, user reviews, and local search Web sites; 3. to establish procedures for defining the context of use to be employed in the recommendation of POIs. The new design should be flexible that can be easily extended to any category of POI. In an Active mode which works like an asynchronized messaging system where a user may proactively seek for answers or advice by depositing a question on the messaging system. The messaging system will automatically disseminate the question which is related to a specific location, to a group of users who are either within the proximity currently or have just recently been there. The users are supposed to help each other; karma or some sort of point/score system can be added to distinguish the most helpful users in the community. A name is suggested for this system called "Lemmings" which stands for Location-based Mutual and Mobile Information Navigation System; it is based on a classical video game where a group of creatures have to work and win through the puzzle game together. A high level design of such location-based collaborative system which pulls information from several sources is shown in Figure 1.

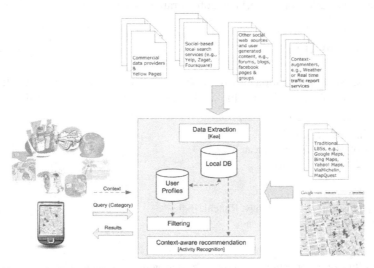

Fig. 1 Abstract design of a location-based collaborative system

2   Background

Our proposed system is extended from a recommender system. Papers related to past & current recommender system are reviewed. In the early days, most research of recommender system is focused on recommender algorithms, like collaborative filtering. After that, some researchers continued working on the algorithmic aspects of recommenders, including a move to hybrid and group recommenders; others have been researching the application of recommenders in specific domains or user interface aspects of recommender systems.

Recommender systems are applied in various applications and websites. They have been deployed on several large e-commerce websites, such as Amazon.com; they are being integrated into corporate document warehouses; and they are still the center of focus for several research groups around the world (Mark van Setten ， Sean McNee and Joseph Konstan, 2005). Moreover, these systems are appearing in products and services used by people around the world, such as personalized television programming and Internet broadcast radio stations, movie recommenders, and even dating services.

2.1   Intelligent Advisor

Quoted from the research paper (Rˇazvan Andonie, J. Edward Russo, Rishi Dean, 2005), intelligent advisor can be known as a recommender system having some key features. For example:

a.) Intelligent advisor can extract knowledge from a customer, build and update corresponding customer profile during each interaction with a customer.
b.) The customer profile built by the intelligent advisor can be further mined.
c.) Intelligent advisor can use behavioral science techniques and create a customer dialog that embraces what/how people think, rather than forcing consumers to feed an optimization algorithm.
d.) Intelligent advisor can continually learning from its interactions with consumers to improve its functionality.
e.) Intelligent advisor can keep working healthily when facing data that are uncertain, noisy, sparse, or missing.
f.) Intelligent advisor should be able to work in real-time to meet the requirements of an Internet application.
g.) Intelligent advisor should be largely domain-independent, and be able to customize the same platform for other applications (e.g., selling computers, cars, financial services).

Intelligent advisor has a stronger learning capacity compared with old recommender system. It can not only learn positive knowledge from a customer and/or the interactions with customers, and also prevent negative data processing from the interactions. It can become more and more talented during the operations.

2.2 Positioning Technology

Positioning is one of the key issues in the proposed system. As a Location based services, apart from the already described technology, require specific infrastructure for positioning the mobile terminal. We need to know the real-time locations of our users in order to achieve the new functions in our system.

The systems offering positioning for mobile terminals in LBS are divided to three main classes: satellite positioning, network-based positioning and, local positioning. Different positioning systems and techniques vary with their features, such as accuracy, reliability and time-to-fix. Each system has its own place, and they complement each other in certain cases. Satellite systems do not work well in deep canyons and indoors where cellular coverage may be denser. Network-based positioning may be more imprecise in rural environments with fewer base stations, where satellite visibility is better.

Satellite positioning systems use an infrastructure of earth-orbiting satellites and receiver terminals. The terminals calculate the position based on the information received through the radio signals from three or more satellites. The terminal-based method provides 10-40 meter accuracy. It renders the user totally independent of the mobile network with respect to positioning, and, in principle, allows access to any location-based services from third-party service providers.

Most well know and widely used satellite positioning system is GPS, operated by US Ministry of Defense. In mobile network environment, there exists also a solution called Assisted-GPS solution, where additional data for the GPS receiver is send through mobile network. The network-based positioning systems refer to positioning methods where the mobile telecommunication networks are used for providing or supporting the positioning of mobile terminals. They include several different methods that are standardized in mobile network specifications. The third main class of positioning systems is local positioning. It refers to positioning that operates only in restricted area and based on short distance signal transmission (Aphrodite Tsalgatidou, Jari Veijalainen, Jouni Markkula, Artem Katasonov and Stathes Hadjiefthymiades).

Some of the researchers started to make the positioning system more talented. The place learning method can be used to predict users' location after some time by learning users' behaviors. This method works with two steps:

Detecting Visits to Shops: The system judges that the user has visited a shop when GPS signals are continuously unavailable for a period of time longer than a threshold. The system then records the location of the visit, and also records the length of time that signals were lost, as the approximate duration of the visit.

Finding Frequented Shops: The actual procedure of our algorithm is as follows. First, for each shop in the database (the system must have a database containing the names and the locations of the shops in the area) the system searches for past visits within a predefined distance from the location of the shop. We call these visits sample visits. Then, by applying two-tailed t-test to the sample visits, the system tries to judge if it is plausible that the shop was frequently visited by the user. In other words, if the assumption that the shop was frequented by the user reasonably explains the past visits recorded nearby that shop, the system judges that the shop was frequently visited by the user. (Yuichiro Takeuchi and Masanori Sugimoto, 2006)

With the positioning system and place learning method, our new system can achieve the functions to detect users' location immediately and accurately and also can predict user's location in the future.

2.3 P2P-Related Problems in Recommendation Systems

Our new system will be a combination of B2C and C2C system. C2C part will be the most important because the main purpose of our system is forcing people to help each other even they are strangers. P2P-related problems in recommendation systems are reviewed from previous articles.

Following are some of P2P-related problems concluded by other researchers:

Dynamic IP addresses: The usage of fixed IP address on the Internet is less common than the dynamic IP address. Dynamic IP addresses make it impossible to directly contact with IP address online.

Lack of trust: Communicating with strangers on the Internet is not as easily as in real world. People cannot see each other, can easily enter and leave rather than in real world. It is much more difficult to trust strangers in digital world than in real world.

Selfish behavior: People may show their worst dimension on the Internet. They want to get more and are willing to give less than in real world. (Johan Pouwelse, Michiel van Slobbe, Jun Wang, Marcel J.T. Reinders& Henk Sips, 2005)

There are still some more problems occur in P2P-related network behaviors. They are needed to be taken care of in designing and operating the system, not only in technology side, but also business and human behavior dimensions.

3 Modular Design of the Lemmings System

In the new Lemmings system, we should have user management module, service supplier module, customizing location module, real-time monitoring module, message module and peer-to-peer connection module. The new Real-time Collaborative Mobile Advisor system will cover the problems occurred in the existing service guide system and achieve more useful functions.

3.1 Real-Time Monitoring Module

In this module, server of the system will collect the locations of all the clients for every minute. So the server can know where the user located in and so that know the user is near which location.

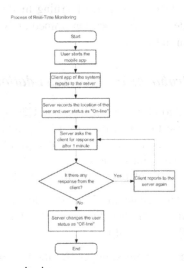

Fig. 2 Process of real-time monitoring

3.2 Service Supplier Module

In this module, the system can store the basic information of the service suppliers, including the category, name, address, telephone, and other useful information of the service suppliers. When the user needs to search for service suppliers, the system can provide the information of service suppliers based on the requirement of the user.

Most functions in this module have already achieved by existing service guide system. We need to improve this module to fit with other modules in our system, e.g. we can recommend some service providers to the user based on his/her real-time location with the real-time monitoring module.

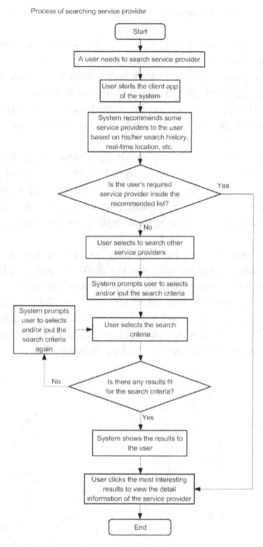

Fig. 3 Process of searching for service providers

3.3   *Message Module*

This is the core module of the new Real-time Collaborative Mobile Advisor system. After the user view the detail information of the selected service provider, he/she may have such questions.

Are there free seats in the restaurant at this moment? If no, how many people are there waiting for seats? How many gifts remained for the 100 gifts per day activity?

User will also have some questions when he/she point a customized location from the map.

Is there a traffic jam near the road I pointed? How is the weather near the area? Is it interesting travelling to that mountain?

Existing service guide system cannot help users to answer these questions. Users can only help themselves, e.g. make phone calls to the service provider, but there may is no response, or go to the location directly, but may be disappointed when there is no seats, activity is finished, there is a traffic jam, weather is bad, etc.

The new system will be able to solve this problem. After the user views the detail information of the selected service provider or point a customized location, the new system allows the user to send message back to the system. The system will check if there are any other users near (e.g. in the area inside 500m distance) the location selected, if yes, the system will push the message to the users searched.

Users near the location will be very easy to answer the questions as listed above. When they are waiting for dishes at the desk or waiting in line for gift, they can help to answer those questions. The answers will be sent back to the server, and then sent back to the user who asked the question. What is more, the system can allow the respondent to send not only text but also image back to the user who asked questions. The user can see the spot situation even as he/she is near the location by the Real-time Collaborative Mobile Advisor system.

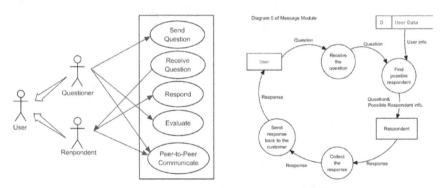

Fig. 4 (a) Use Case diagram and (b) Flow diagram of the message module

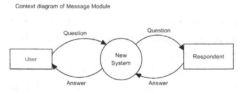

Fig. 5 Context diagram of the message module

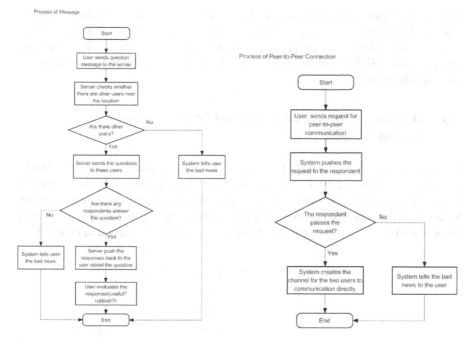

Fig. 6 Process diagram of the message module

Fig. 7 Process diagram of the peer-to-peer connection module

3.4 Peer-to-Peer Connection Module

After the user asks the question and receive response from other users, the system can allow peer-to-peer connections between the users so that they can communication with the things they are both interested more conveniently and effectively. What is more, users can make friends by using this system!

First the user who asks the question should send a request through the server to the respondent, if the respondent passes the request; they can make peer-to-peer communication with each other. Their information will also be stored into the others' account so that they can easy to get in touch next time.

3.5 User Management Module

This module can help the new system to manage the users and achieve the functions mentioned before. Guest users can use our system to search service providers without registration. But asking questions for help and respond questions to help others requires a user account. With the users' accounts, the system will become easier to manage the message sending and receiving, peer-to-peer connection, and also can give some gifts (e.g. coupon of some service providers) to encourage users to response the questions so that help other users.

This module can also collaborate with real-time monitoring module, when the system detects the user near a service provider, the system can push some advertisement of the service provider to them. It is much better than pushing advertisement to all users because for most of the users, the advertisement far away has no value.

3.6 Database Design and Technical Implementation Considerations

A central server with a database is needed to handle the client applications. Inside the database, there are two main entities, user and business suppliers. Besides basic information of the user and business, we also store the l the real-time location of the users and business suppliers inside the database, so that we can know which users are near a particular business supplier at a particular moment.

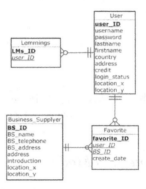

Fig. 8 Database schema

MySQL is a widely-used open source database server. It has good compatibility, stability, security and usability. It can be easily operated by most server engine and client applications in different language. We need to develop applications compatible at least three platforms, Android, Apple OS and Windows Phone. In Android platform, Java is the most compatible to develop applications, while in IOS and Windows Phone platforms, C and C# are the best correspondingly. Because the different mobile client applications in different platform, we also need different Server side platforms. Apache with PHP and serve C and C# well for Apple OS and Windows Phone platforms. Tomcat with JSP can serve Java client well for Android platform.

4 Conclusion

A new designed system called Lemmings which is based on collaborative location-based services will play a positive role in the online society. It encourages users help each other and trust each other each for information inquiry of daily matters. In this paper, the design of Lemmings is presented in the form of software analysis and design diagrams. Future work includes testing the prototype and the efficiency of target message sending algorithms.

References

van Setten, M., McNee, S., Konstan, J.: A Workshop on the Next Stage of Recommender Systems Research (2005)

Andonie, R., Edward Russo, J., Dean, R.: Crossing the Rubicon for An Intelligent Advisor (2005)

Samsung Official Website samsung.com, Detail of Samsung Galaxy S4

Wikipedia wikipedia.org, Smartphone Wiki

Android Official Website, android.com, latest release version of Android

Apple Official Website, apple.com, latest release version of iOS

Tsalgatidou, A., Veijalainen, J., Markkula, J., Katasonov, A., Hadjiefthymiades, S.: Mobile E-Commerce and Location-Based Technology and Requirements

Takeuchi, Y., Sugimoto, M.: CityVoyager: An Outdoor Recommendation System Based on User Location History (2005)

Pouwelse, J., van Slobbe, M., Wang, J., Reinders, M.J.T., Sips, H.: P2Pbased PVR Recommendation using Friends, Taste Buddies and Superpeers (2005)

Tsalgatidou, A., Veijalainen, J., Markkula, J., Katasonov, A., Hadjiefthymiades, S.: Mobile E-Commerce and Location-Based Technology and Requirements

Neural Network Based Early Warning System for an Emerging Blackout in Smart Grid Power Networks

Sudha Gupta, Faruk Kazi, Sushama Wagh, and Ruta Kambli

Abstract. Worldwide power blackouts have attracted great attention of researchers towards early warning techniques for cascading failure in power grid. The key issue is how to analyse, predict and control cascading failures in advance and prevent system against emerging blackouts. This paper proposes a model which analyse power flow of the grid and predict cascade failure in advance with the integration of Artificial Neural Network (ANN) machine learning tool. The Key contribution of this paper is to introduce machine learning concept in early warning system for cascade failure analysis and prediction. Integration of power flow analysis with ANN machine learning tool has a potential to make present system more reliable which can prevent the grid against blackouts. An IEEE 30 bus test bed system has been modeled in powerworld and used in this paper for preparation of historical blackout data and validation of proposed model. The proposed model is a step towards realizing smart grid via intelligent ANN prediction technique.

1 Introduction

Power grid is a complex interconnected networks in which tripping of one line overlodes other lines and initiate cascading failures. Indian power grid recently experienced two consecutive severe blackouts on July 30th and 31st, 2012. It was spread across 22 states in Northern, Eastern, and Northeast India and affected over 600 million people [1]. In such a scenario the question arises is, whether the prediction of similar type of blackouts is possible on the basis of analysis of past blackout events.

The analysis of cascading failure in the grid needs power flow model which can analysed the grid in steady state as well as under dynamic state. The Power flow model fundamentally classify as deterministic and probabilistic model. A Probabilistic model was proposed by [2], and [3], for stability of power in the power

Sudha Gupta · Faruk Kazi · Sushama Wagh · Ruta Kambli
COE-CNDS VJTI, Matunga, Mumbai, India 400019
e-mail: sudhagupta@somaiya.edu

© Springer International Publishing Switzerland 2015 173
R. Buyya and S.M. Thampi (eds.), *Intelligent Distributed Computing*,
Advances in Intelligent Systems and Computing 321, DOI: 10.1007/978-3-319-11227-5_16

system in terms of density function. The paper compares the probabilistic and deterministic results of line contingency. Simulation results showed the effect of changes in nodal data with respect to power flow in the transmission lines. The network uncertainties in power system was modeled in [4] and [5] where load flow analysed with consideration of network topology uncertainties .

The probabilistic framework for power grid cascade failure analysis was explained in [6] along with statistical decision theory. Simulation result showed that in normal grid operation probability distribution was the normal (Gaussian), whereas in cascading failure distribution was no more Gaussian. Hence, probabilistic load flows are useful for the analysis of variations in power flow in transmission line over a period of time.

The statistical behavior of grid under normal and blackout condition can be used for generation of control signals for protective relays in early warning system using machine learning tools like artificial neural network (ANN).

An ANN is a machine learning tool which map and model large class of complex and nonlinear phenomena directly from the inputs and used as a pattern recognizer or classifier to predict events. An overview of application of neural network in power system controller design and in load forecasting was given in [7].

An adaptive pattern recognition approach using artificial neural networks (ANN) used by [8] for security assessment in electric power system. ANN based multi-layer prediction model with contingency ranking was used in [9] for steady state, transient and dynamic security assessment of the Power systems. The performance of various ANN controllers was compared by [11].

A Hybrid feedforword/feedback ANN used in electric power industry to access transient stability and electric load forecasting [12]. The multi-layer feedforward network was first introduced in [13] for power system applications. The ability of ANN to learn, generalize and pattern classifiers was used in [14] for fault detection in a transmission line power system.

The aim of this paper is to provide an ANN based cascade prediction system with probabilistic load flow analysis for early warning of cascade failure in smart grid through case study, modeling, and simulation. The paper is organized as follows: Section 2 describes probabilistic load flow and cascading failure. Neural Network concept is given in Section 3. Proposed model is explained in Section 4. Section 5 is a case study using IEEE 30 bus test bed system to validate proposed model. The paper is concluded in Section 6 along with future research direction.

2   Power Flow and Cascading Analysis

The power flow analysis is necessary to plan ahead and account for various possible situations (steady state and dynamic state) of power system. The analysis in normal steady-state operation is called a power flow study [15] and it targets on determining the voltages, currents, real and reactive power flows in a system under a given load conditions. The active power equations of a power flow [15] are given as,

$$S_i = P_i + jQ_i = (P_{Gi} - P_{Di}) + j(Q_{Gi} - Q_{Di}) \tag{1}$$

where the complex power supplied by the generator is given by

$$S_{Gi} = P_{Gi} + jQ_{Gi} \tag{2}$$

In (1),and (2), P, and Q are real and reactive power of the system. The complex power, drawn by the load is

$$S_{Di} = P_{Di} + jQ_{Di} \tag{3}$$

The real and reactive powers injected into the ith bus (node) are

$$P_i = (P_{Gi} - P_{Di}); \quad i = 1, 2, 3, \cdots, n \tag{4}$$

$$Q_i = (Q_{Gi} - Q_{Di}) \tag{5}$$

The reactive power of the system is non-linear and complex hence computational complexity is more at the same time the real power flow is linear and simple to solve system state with fast load flow computation after adding or tripping a line.

Assume the net power injected into a node is real and equal to the total amount of power flowing to neighboring nodes through links (transmission lines or transformers) that is

$$P_{ij} = \frac{j(\theta_i - \theta_j)}{X_{ij}} \tag{6}$$

where θ_i, θ_j is the voltage phase angles at node i and j and X_{ij} is the series reactance of the link between nodes i and j.

2.1 Probabilistic Load Flow

Factors which can cause cascading failure are load fluctuation, random failure of generator unit and random failure of transmission and distribution component. The variables in power grid like node voltage, current and power flow in transmission line are treated as random variables in probabilistic load flow model and can be defined by normal distribution [15].

The states of power system (states of transmission line) has been described in the form of two states, normal state and failure state, hence, transmission lines defined as a two state stochastic process that develops in time and controlled by probabilistic laws as shown in Table 1.
where:

C = Transmission capacity of load line

q = Probability of failure (80 % of IC)

$(1 - q)$ = Available capacity

The Probabilistic laws for cascade failure analysis consider power loading of each transmission line as a random number with uniform distribution U (0 to 1) then compared this random number with forced outage rate (80% of installation Capacity

Table 1 Outage Table of Transmission Line

Avalable capacity	Outage capacity	Deterministic probability	Cumulative Probability
C	0	1-q	1
0	C	q	q

(IC)) q, which is used further to determine whether transmission line state is in failure or running state.

Consider outage capacity $\bar{X}$ is a random variable and its probability is $P_{(n-1)}(\bar{X})$.

For a forced outage rate q_n, and available capacity C_n, the new probability $P_{(n)}(\bar{X})$ of the outage capacity can be calculated by using convolution formula

$$p_n(\bar{X}) = p_{n-1}(\bar{X})p(0) + p_{n-1}(\bar{X} - c_n)p(c_n) \tag{7}$$

and the two state probability of the nth transmission line can be calculated by

$$p(c_n) = q_n; \quad p(0) = 1 - q_n \tag{8}$$

Equation (8) is useful for state prediction of power system.

The power flow in transmission line under normal working of the grid probabistically represented by normal (symmetrical) distribution in this paper to calculate and analysis of load duration curve, mean, variance and higher order moments.

Let, $n = (1, 2, \cdots, N)$ are a number of samples in normal distribution under normal working of the grid. Under cascading condition (disturbance) the probability distribution of the grid power may not be symmetrical (Normal/ Gaussian). The behavior of cascading link failure statistically analysed in this paper using distinct data peak near the mean the higher order moments namely third moment and forth moment of probability distribution.

The Probability Density Function (PDF) and Cumulative distribution function (CDF) is used to observe time scale (sample) for prediction of next line contingency. The value of CDF becomes 1 at the time of blackout. The probabilistic model is used in this paper to prepare normal and blackout historical data base. The mean variance and higher order moments of normal and blackout case has been used for training of ANN model.

3 Neural Network Architecture

The ability to map and model large class of complex and nonlinear phenomena directly from input, makes neural networks one of the powerful pattern recognition/classification machine learning technique. As shown in Fig.1 all of the processing that goes on within each processing element must be completely local i.e., it must depend only upon the current values of the input signals arriving at the

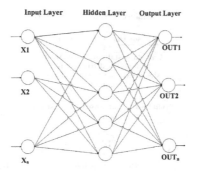

Fig. 1 Artificial Neural Network

processing element via impinging connections and upon values stored in the processing element (local memory).

Let,

$X = (x_1, x_2, \cdots, x_i, \cdots, x_n)$ is the input feature vectors. The net input is calculated as

$$y_{ni-j} = b_j + \sum_{i=1}^{n}(x_i w_{ij}) \tag{9}$$

Where,

w_{ij} = weights in a matrix form

y_{ni-j} = net input to output unit yj

b_j = bias for output layer

The sum of the weighted input signal (9) is applied with an activation function to obtain the response of ANN. A feedforward supervised ANN has been used in this paper for blackout prediction. In a feedforward ANN, connections between the neurons do not form a directed cycle and information moves only in forward direction. In supervised learning algorithm very popular algorithm is backpropagation. In which, activation function should be differentiable [16].

The standard backpropagation algorithm often gives undesirable outputs [17] on large-scale problems and depends on the user dependent parameters, learning rate and momentum constant. The Scaled Conjugate Gradient (SCG) method avoids the line-search per learning iteration. It involves two independent steps to find next point in the process. Finding the search direction is the first step i.e, determining the direction in weight space in which new current point is to be searched. After the first step of finding the direction, step size has to be determined to find how far to go in that direction. As shown in Fig.1 Each processing element in ANN has a single output connection which branches into as many collateral connections as desired and each carrying the same signal (processing element output signal). The processing element output signal can be of any mathematical type desired.

The convergence criteria decides the speed-up of process. that is, reduction in error is directly proportional to speed-up. Line-search per learning iteration, which makes other second order algorithms slower, that is avoided in the SCG, by using a step size scaling mechanism. Hence, this paper uses SCG supervised learning algorithm to train ANN.

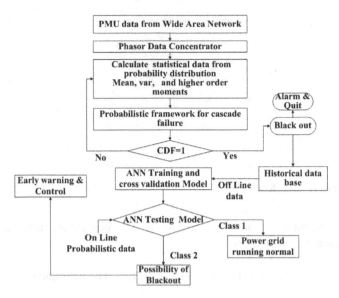

Fig. 2 Proposed mode for blackout prediction

4 Proposed Cascade Prediction Model

A model of an early warning system for prediction of emerging blackout in power grid using ANN, proposed in Fig. 2. As shown in Fig. 2 PMU (Phasor Measurement Unit) of wide area network collects the grid data (% power loading on transmission line) and sends to phasor data concentrator (PDC) for processing. At PDC it is first converted into probability distribution (Normal/Gaussian, distribution) and from the distribution curves the variables such as mean, variance, and higher order moments have been calculated. A historical data base of normal grid operation as well as blackout case are generated from these variables using probabilistic framework [6] of cascade failure. As shown in Fig. 2 The higher order moments of various normal and blackout cases are calculated and stored in historical database.

The proposed model requires these historical data as an input to train and validate ANN for future blackouts prediction. Predicted output is further given to system protection center for emergency action like system protective relays, breaker or forced outage for system islanding.

5 Case Study for Validation of Proposed Model

An IEEE 30 bus test bed model has been used for cascade failure analysis and generation of historical database. As shown in Fig.3, test bed system having 30 buses, out of which 6 are generator buses and 21 load buses with 42 transmission lines having 289.1 MW generation and 283.4 MW load capacity.

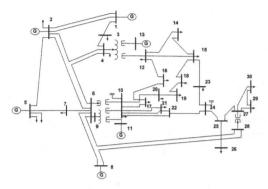

Fig. 3 IEEE 30 bus test bed system

5.1 Probabilistic Model

The test bus system has been modeled probabilistically using normal distribution as shown in Fig.4 (blue curve). For the analysis of cascading failure, a random line L21 between nodes 10 to 21 from Fig. 3 has been tripped . Tripping of this line redistributes power in the neighboring lines , hence, changed probability distribution according to Table 1. Probability distribution of the test bus system under normal operating condition and under cascading [6] is shown in Fig 4. The probability of error rate (cumulative hazards) plot in terms of percentage line loading is shown in Fig.5

The highest probability of next line tripping is the line which is having highest PDF. This scenario initiated cascading failure and finally landed up to blackout (CDF=1). The same is reflected in PDF plot, Fig. 4 and CDF plots, in Fig. 5, and Fig 6. Hazard rates are used here to estimate disturbed variables and to quantify the effect of predictor variables. It represents the effects of explanatory and disturbed

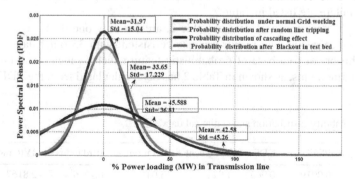

Fig. 4 Probability distribution of power flow in IEEE 30 bus test bed system from normal grid operation to blackout

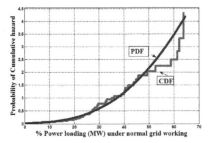

Fig. 5 Cumulative hazards under Normal working of grid

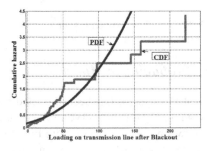

Fig. 6 Cumulative Hazards after blackout

variables. Fig 5 is a plot of baseline hazard function, $h_o(t)$. For a base line relative to $h_o(t)$ (Normal), the hazard regeneration (cascading) can be calculated as,

$$h_x(t) = h_0(t)e^{\sum_{i=1}^{N}(x_i b_i)} \tag{10}$$

Where, x is a input sample , $h_x(t)$ is the hazard rate (cascade failure rate) at x, and b is a estimation coefficient. As shown in Fig.5 the baseline hazard function depends on time, t, but the predictor variables (disturbed variable) after cascading in Fig. 6, do not depend on time. This distributed PDF and CDF plots has given very distinct higher order moments. The variables under normal grid operation and under cascading failure are listed in Table 2.

The graphs and statistical data of Table 2, are for only one data set of normal and blackout case, however results are also consistent with 35 other data sets we tested. A historical data base of probabilistic feature such as mean, variance and higher order moments, as shown in Table 2 has been calculated from the PDF, CDF

Table 2 Probability distribution and statistical data

Grid operation	PDF	CDF	1st moment	IInd moment	IIIrd moment	1Vth moment
Normal	0.0019	0.960	31.97	15.04	0.2286	2.8167
Cascading	7.2488e-04	0.994	33.65	17.229	0.4856	2.3796
Cascading	3.5912e-05	0.998	45.588	36.81	2.2861	9.6091
Blackout	6.6791e-05	0.999	42.58	45.26	2.6515	10.425

plots of Figs. 4, 5, and 6 for normal and cascade operation from the test bench. A historical database consists of 35 set of such cases have been used for training and testing of neural network.

5.2   Neural Training and Testing Model

ANN has been trained using four feature vector (as shown in Table 2) from historical database. 75% of data has been used for training 15% for validation and 10% data has been used for testing purpose using feed forward network and, SCG fast supervised learning algorithm. As shown in Fig. 7 where w is a weight vector and b is a bias term.

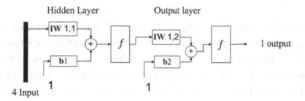

Fig. 7 ANN training model

Table 3 is a model parameters, used in training of ANN where default value sigma determine changes in weight for second derivative approximation and Lambda is a regulation parameter. Fig.8 is a plot between no. of iteration and the mean square error between targeted (actual) and predicted output. Mean Square Error (MSE) plots in Fig. 8 clearly shows that the validation and test curve are very close to each other, hence, there is no chance of over-fitting. As shown in MSE graph the best validation performance is $1.3371e^{-06}$ at 22 epochs. The predicted output of ANN is used as an input to early warning system for further preventive actions.

Table 3 ANN training parameters

No.of input features	4
No. of layers	2
No. of outputs	1
Hidden layer size	10
Activation function	Radial
Minimum gradient	1e-06
Maxmum validation	06
Sigma	5.0 e-5
Lambda	5.0 e-7

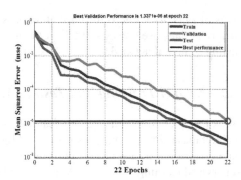

Fig. 8 ANN training, testing, and validation graph

6 Conclusions

The paper presented cascade failure prediction model with the advancement of ANN tool. In the case study simulation results and statistical analysis shows that under normal power flow the probability distribution observed was normal but as cascading failure propagates distribution curves shifts towards heavy tail. Hence, we got very distinct changes in higher order moments in cascading case and this helped to select feature vectors for training of ANN model for cascade prediction. It was observed that including higher order moments mean square error between actual and predicted output reduces significantly which increases accuracy of validation and prediction model in a very significant way. ANN concept can be further extended for prediction and identification of critical link in the grid. Machine learning concept in power system still needs research to demonstrate its practicality on real time simulation.

Acknowledgment. We acknowledge funding from WorldBank under TEQIP, subcomponent 1.2.1 through the Center of Excellence in Complex and Nonlinear Dynamical Systems (CNDS), and coordinator VJTI, Matunga Mumbai, India. Special thanks to Dr. N.M. Singh for technical support and guidance.

References

1. A Report of The enquiry committee on grid disturbance in Northen region on 30th July 2012 and in Northern, Eastern and North-Eastern region on 31 st July 2012, 16th August 2012 New Delhi
2. Borkowska, S.B.: Probabilistic load flow. IEEE Trans. on Power Apparatus and Systems PAS-93(3), 752–755 (1974)
3. Allan, R.N., Borkowska, B., Grigg, C.H.: Probabilistic analysis of power flows. In: Proceedings of the Institution of Electrical Engineers, London, vol. 121(12), pp. 1551–1556 (December 1974)
4. Min, L., Zhang, P.: Probabilistic load flow with consideration of network topology uncertainties. In: The 14th International Conference on Intelligent System Applications to Power Systems, ISAP 2007, Kaohsiung, Taiwan, November 4-8, pp. 7–11 (2007)

5. Zhang, P., Lee, S.T.: Load flow computation using the method of combined cumulants and Gram-Charlier expansion. IEEE Trans. on Power Systems 19(1), 676–682 (2004)
6. Gupta, S.R., Kazi, F.S., Wagh, S.R., Singh, N.M.: Probabilistic framework for evaluation of smart grid resilience of cascade failure. In: IEEE Innovative Smart Grid Technologies Conference (ISGT) Asia, Kuala Lumpur, Malaysia, May 20-23, pp. 264–269 (2014)
7. Guo, C.-X., Jiang, Q.-Y., Cao, X., Cao, Y.-J.: Recent developments on applications of neural networks to power systems operation and control: An overview. In: Yin, F.-L., Wang, J., Guo, C. (eds.) ISNN 2004. LNCS, vol. 3174, pp. 188–193. Springer, Heidelberg (2004)
8. Sobajic, D.J., Pao, Y.-H.: Artificial neural -netbased dynamic security. IEEE Trans. on Power Systems 4(1), 220–228 (1989)
9. Aggoune, M., El-Sharkawi, M.A., Park, D.C., Damborg, M.J., Marks, R.J.: Preliminary results on using artificial neural networks for security assessment. IEEE Trans. on Power Systems 6(2), 890–896 (1991)
10. Swamp, K.S., Prasad Reddy, K.V.: Neural Network based Pattern Recognition for Power System Security Assessment. In: CISIP 2005 (2005)
11. Hassan, L.H., Moghavvemi, M.: Current state of neural networks applications in power system monitoring and control. International Journal of Electrical Power and Energy Systems 51, 134–144 (2013)
12. Bourguet, R.E., Antsaklis, P.J.: Artificial Neural Networks In Electric Power Industry. Technical Report of the ISIS (Interdisciplinary Studies of Intelligent Systems) Group, No. ISIS-94-007, Univ of Notre Dame (April 1994)
13. Wong, K.P.: Artificial intelligence and neural network applications in power systems. In: 2nd International Conference on Advances in Power System Control, Operation and Management, December 7-10, vol. 1, pp. 37–46 (1993)
14. Bashier, E., Tayeb, M.: Faults Detection in Power Systems Using Artificial Neural Network. American Journal of Engineering Research (AJER) 2(6), 69–75 (2013)
15. Sullivan, R.L., Lee, R.: Power system planning. McGraw-Hill International Book Company
16. Rumelhart, D.E., Hinton, G.E., Williams, R.J.: Learning representations by back-propagating errors. Nature 323(6088), 533–536 (1986)
17. Moller, M.F.: A Scaled conjugate gradient algorithm for fast supervised learning. Neural Networks 6(4), 525–533 (1993)

Hybrid Genetic Fuzzy Rule Based Inference Engine to Detect Intrusion in Networks

Kriti Chadha and Sushma Jain

Abstract. With the drastic increase in internet usage, various categories of attacks have also evolved. Conventional intrusion detection techniques to counter these attacks have failed and thus substantial systems are needed to eliminate these attacks before they inflict huge damage. With the ability of computational intelligence systems to adapt, exhibit fault tolerance, high computational speed and error resilience against noisy information, a hybrid genetic fuzzy rule based inference engine has been designed in this paper. The fuzzy logic constructs precise and flexible patterns while the genetic algorithm based on evolutionary computation helps in attaining an optimal solution, thus their collaboration will increase the robustness of intrusion detection system. The proposed network intrusion detection system will be able to classify normal behavior as well as anomalies in the network. Detailed analysis has been done on DARPA-KDD99 dataset to specify the behavior of each connection.

1 Introduction

Internet is a global public network. On one side it has an enormous potential to provide open and easy communication and reach the end users, while on the other hand it brings a lot of risks with serious threats, intrusions and vulnerabilities to the systems. Set of operations that endeavor to subvert the integrity, confidentiality and availability of a resource is defined as intrusion [1]. Thus, an intrusion detection system is a security system that acts as a watchdog for computer systems and network traffic and determines the entities that try to disrupt the security mechanism [2].

Kriti Chadha · Sushma Jain
Thapar University, Patiala, India
email: kritichadha1989@gmail.com, sjain@thapar.edu

© Springer International Publishing Switzerland 2015   185
R. Buyya and S.M. Thampi (eds.), *Intelligent Distributed Computing*,
Advances in Intelligent Systems and Computing 321, DOI: 10.1007/978-3-319-11227-5_17

The intrusion detection systems (IDS) are mainly of two types:

- Host Based IDS (HIDS)-HIDS attempts to identify anomalous or unauthorized behaviour on a specific device on the network. Here an agent is installed on each system which monitors operating system, write data to log files or trigger alarms.
- Network based IDS (NIDS)-NIDS are placed at strategic points within the network to monitor traffic moving to and from all the devices on the network. They work in promiscuous mode and collects packets using either a network tap or hub. The system then processes these packets and flags/triggers alarms.

The intrusion detection system use one the two detection methods:

- Misuse detection-Also known as signature based detection; it detects intrusions by referring to database containing information about system vulnerabilities and previous attack profiles. Hence they have potential of very low false rates and are more accurate in detecting attacks. But its drawback is difficulty in maintaining and updating the database with the latest system vulnerabilities.
- Anomaly detection- They reference a baseline to detect deviation from normal or expected behaviour of system or users. Thus they can attempt to detect new or unforeseen vulnerabilities. But the disadvantage is high false alarm rate as the entire behaviour of the system is not covered

The proposed methodology aims to use network based misuse detection scheme to design an IDS based on hybrid genetic fuzzy rule based inference engine.

A fuzzy rule based system is based on fuzzy if-then rules and has been mainly used in classification field [3].

A fuzzy rule based system is divided into mainly three conceptual components:

- Fuzzy rule base-which consist of set of fuzzy rules and the linguistic labels represent the input variables.
- Database (Dictionary)-which specifies the number of linguistic labels for each variable and the membership function of the fuzzy set concerned with the linguistic terms.
- Fuzzy reasoning method-which derives conclusion from the given fuzzy if then rules and facts.

A grid type fuzzy partition is formed where each input is divided into antecedent fuzzy sets with linguistic labels. Confidence is calculated corresponding to each fuzzy if then rule to define how effectively a fuzzy rule identifier assigns an input pattern to a class label [5].Thus two main goals are achieved i.e. accuracy and interpretability of fuzzy rule based system [4].

Genetic algorithms (GA) are adaptive heuristic search algorithms inspired from Darwin's theory of evolution which is based on natural selection and evolution and therefore these algorithms are very successful in search and optimization problems. Genetic algorithms begin with the population of randomly generate chromosomes, which are candidates of the problem. A fitness function is assigned to each chromosome to determine the adaption of solution represented by that chromosome. During successive generations, fitness of each chromosome is evaluated and on basis of them, new population is brought forth using different selection mechanisms and genetic operators such as crossover and mutation. Thus information is accumulated from unknown search space and optimized solutions are generated through competition and controlled variation.

Three genetic fuzzy rule based approaches have been already devised for the learning process [8]:

- Michigan approach-Here each chromosome represents one rule.
- Pittsburgh approach-Here a whole set of rules is encoded as an individual/chromosome.
- Iterative rule learning-Here each chromosome represents one rule but selects only the best solution and discards the remaining solution.

In the proposed work, a different approach is adopted from the existing three approaches where genetic fuzzy rule based system generates a set of rules and then a mathematical model is applied to select the best rules from the already existing pool of rules.

The rules will be formed on the basis of 41 attributes in each connection of the KDD-99 dataset. Then they will be evaluated and optimized to obtain a set of rules which will help in detection of anomalies and the normal behaviour of the connection in the network.

2   Related Works

The first intrusion detection model was given by Denning in 1987[6]. Since then many intrusion detection models have been constructed to determine the behavior of the network accurately and in an efficient manner. The IDS has been classified on the basis of various characteristics as given by [7].

In 1999, KDD99 dataset was derived from the DARPA98 network traffic dataset by ACM SIG-KDD International Conference on Knowledge Discovery and Data Mining. It consists of TCP connections and is the most widely used dataset for network intrusion detection and has training and testing datasets and each connection consists of 41 attributes.

Subsequently various artificial intelligence, machine learning and computational techniques have been applied on various intrusion detection models to obtain precise results while confronting various problems such as vast network traffic, noisy information, continuous adaptation to changing environment.

(Liao and Vumeri, 2002)[10] proposed k-nearest neighbor method for feature selection in KDD99 dataset. (Middlemiss and Dick, 2003) [9] used KNN in conjunction with genetic algorithm to classify the normal and abnormal behavior. The

41 attributes rendered by KDD99 dataset were ranked by (Mukkamala and Sung, 2002)[14] through support vector machines and they again combined SVM and neural networks in [15] to rank the features. Feature selection based on rough set was introduced by [12].

(Stein *et al.*, 2005) [13] utilized decision tree classifiers with genetic algorithm in intrusion detection approach while (Hofman *et al.*,2004)[16] gave evolutionary wrapper approach (combining radial basis function with genetic algorithm). Genetic algorithm along with naïve bayes classification was used by (Lee *et al.*, 2006) [11] to generate optimized results for classification on KDD99 dataset. (Lu *et al.*, 2005)[17] applied genetic algorithm to decide the number of clusters based upon Gaussian Mixture Models.

(Gomez *et al.*, 2002)[18] first demonstrated the work of fuzzy classifiers on intrusion detection system. Various GA models such as Michigan, Pittsburgh and Iterative rule learning approaches along with different swarm intelligence techniques have been applied by [19].

In the proposed approach, hybrid genetic fuzzy rule based system is projected utilizing the technique by [20, 21] which is practiced in high dimensional problems to establish a network intrusion detection system which distinctly classify normal class and the different types of attacks prevalent in KDD99 dataset.

3   Fuzzy Rule Based System

In the proposed work, there are mainly three components- fuzzy rule based system, genetic algorithm and computational model. In this section the fuzzy rule based system to detect intrusions is discussed.

The 10% KDD-99 dataset consists of 4, 94,021 records in training dataset. Here 750 random samples are taken into account and there are 41 attributes in each record. Thus in a fuzzy rule based system, there are m i.e. 750 labeled patterns with n=41 dimensionality, i.e. each pattern is represented as

$$x_p = \{x_{p1}, x_{p2}, x_{p3}, x_{p4}, x_{p5}, \ldots \ldots \ldots \ldots \ldots, x_{p38}, x_{p39}, x_{p40}, x_{p41}\}$$

where p=1,2....,m training patterns and each pattern is required to be classified into c classes i.e. 5 classes. Here only continuous attributes have been considered for further processing.

Since the space of pattern taken into account is $[0,1]^n$, so the attribute values of each pattern is normalized in the range $[0,1]$ i.e. $x_{pi} \in [0,1]$.

To obtain the fuzzy rules we perform the following steps:-

1. Five linguistic variables (L_i) (fig. 1) are assigned for each feature considered (i.e. Very Low, Low, Medium, High, and Very High) and each feature definition interval has been uniformly partitioned using triangular fuzzy sets. The membership function of a jth fuzzy set is determined by the following formula:-

$$\mu_j(x) = \left\{ 0,1 - \frac{|x - x_j|}{v} \right. \tag{1}$$

where $x_j = \dfrac{j-1}{L-1}$ and j=1,2...L.

and $v = \dfrac{1}{L-1}$.

2. The compatibility of each training pattern x_j is calculated with the fuzzy if then rule R_j by the following formula:-

$$\mu_j(x_p) = \prod_{i=1}^{n} \mu_{ji}(x_{pi})$$

p=1,2,....m   (2)

where $\mu_{ji}(x_{pi})$ is the membership function of the ith attribute of pth pattern.

3. Fuzzy if then rule is selected according to the given training patterns. The consequent class C_j is calculated as follows:

$$\beta_{Classh}(R_j) = \max_{h=1,2,..,c}(\beta_{Classh}(R_j)) \tag{3}$$

where $$\beta_{Classh}(R_j) = \sum_{x \in Classh} \frac{\mu_j(x_p)}{N_{Classh}} \tag{4}$$

and $\beta_{Classh}(R_j)$ represents the mean sum of compatibility grades in Class h with fuzzy if then rule and N_{Classh} is the number of training patterns taken into consideration corresponding to each $Classh$ and h ranges from 1,2,.....,c. where c=5 i.e. there are mainly five classes-normal, dos attacks, u2r attacks, r2l attacks and probe attacks.

The maximum of $\beta_{Classh}(R_j)$ is evaluated and the one with the maximum value is considered to be the class of that fuzzy if then rule R_j. If the maximum value comes out to be true for more than one class, then the consequent class C_j cannot be determined uniquely and is taken as φ and the corresponding rule is rejected.

4. The confidence or certainty degree refers to how precisely an input pattern classified by the fuzzy rule based system. It is determined by the following:-

$$CD_j = \frac{\beta_{Classh_j}(R_j) - \bar{\beta}}{\sum_{h=1}^{c} \beta_{Classh_j}(R_j)} \tag{5}$$

where
$$\bar{\beta} = \frac{\sum_{h \neq h_j} \beta_{Classh}(R_j)}{c - 1} \tag{6}$$

If the consequent class C_j is φ, then the confidence is also φ. The certainty degree lies in the unit interval [0, 1].The value of CD_j =1 represents very high confidence.

Now to classify an unlabelled pattern, $x_p = \{x_{p1}, x_{p2}, \ldots, x_{pn}\}$, the input pattern with the rule set is matched and the following approach is applied.

1. The class of the unknown pattern is determined by the following equations:-

$$C = \max_{h=1,2..c}(\tau_h) \tag{7}$$

where
$$\tau_h = \max_{\substack{j=1,2..,N \\ C_j \in h}} \{\mu_{A_j}(x_p).CD_j\} \tag{8}$$

2. Thus the successful rule is the one which has highest τ_h but if more than one class has highest τ_h or if $\tau_h = 0$ then the rule is rejected.

Therefore, the fuzzy if then rules are generated in the following manner:-

Rule R_j=If x_1 is A_{j1} and x_2 is A_{j2} and ...., x_n is A_{jn}, then the class is C_j with CD_j, j=1,2...N

where R_j is the label of the jth fuzzy if then rule, A_{j1}, A_{j2},..., A_{jn} denotes the antecedent fuzzy sets, C_j represents the consequent class, CD_j refers to the certainty degree or the confidence in the class label and N is the number of rules.

For example- If protocol is 1.0,..........,and dst_host_srv_count is 1.0,........,and dst_host_srv_diff_host_rate is 0.92,.........and dst_host_srv_rerror_rate is 1.0 ,then the class is r2l attack with certainty degree=1.0.

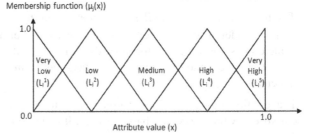

Fig. 1 Triangular membership functions of fuzzy sets

4   Hybrid Genetic Algorithm to Generate Classification Rules

Genetic algorithms are inspired from Darwin's theory of Evolution-"the survival of fittest". This algorithm exploits the historical information in order to get optimized results.

1. Before applying a genetic algorithm, the potential solutions are required to be encoded. The most commonly used encoding scheme is Binary Encoding. Let the population size be N_{pop}, i.e., the number of fuzzy if then rules to be generated. In GA, each individual of the population is represented by chromosome. Here each chromosome represents a rule for a specific class. Since each feature or attribute is divided into five linguistic labels, so the length of the chromosome is equal to the product of the number of features and the number of linguistic labels.

   Each chromosome is a set of genes. Here the first five genes represent first feature, next five second feature and so on (fig.2).
   Each rule is expressed as:-

   R_j= If x1 is l_1^2 or l_1^5, x2 is l_2^1 and x3 is l_1^3, then the class is C_j with CD_j.

L_1^1	L_1^2	L_1^3	L_1^4	L_1^5	L_2^1	L_2^2	L_2^3	L_2^4	L_2^5	L_3^1	L_3^2	L_3^3	L_3^4	L_3^5
0	1	0	0	1	1	0	0	0	0	0	0	1	0	0

Fig. 2 Sample chromosome structure

2. Now, the fitness function is evaluated for each fuzzy if then rule by classifying all the training patterns with the N_{pop} fuzzy if then rules generated. The formula of the fitness function is given as[21]:-

$$Fitness = w_1 \times CD_j + w_2 \times (1 - \text{var}_N) + w_3 \times (1 - lab_N) + w_4 \times Rule_j \quad (9)$$

where CD_j represents the confidence for each fuzzy if-then rule, var_N are the normalized values of the number of features(min=1 and max=n_v), lab_N denotes the normalized value of linguistic labels used. (Here min.value of lab_N is 1 and max. value is $n_v*(l_i-1)$.

$Rule_j$ is given as:-

$$Rule_j = \frac{n_j}{N} \quad (10)$$

where n_j is the number of samples covered by the rule

N is the total number of training samples.

w_1, w_2, w_3 and w_4 represents the weights and $w_1 + w_2 + w_3 + w_4 = 1$.

For example-If protocol is l_1^1,..........,and dst_host_same_srv_rate is l_{33}^5,..........,and dst_host_srv_diff_host_rate is l_{37}^2or l_{37}^3,.............and dst_host_srv_rerror_rate is l_{41}^1,then the class is r2l attack with certainty degree=1.0 and fitness=0.5013.

3. Now, a steady state selection mechanism is applied to get top 10% of the population which has high fitness value. On the rest of the population, rank based roulette wheel selection mechanism is applied.

   In the rank selection method, the population is sorted according to their fitness value; the worst chromosome is ranked 1 and the best with rank N. Here, rank is calculated by the following formula:-

$$P_{rank}(i) = \frac{2-s}{\mu} + \frac{2i(s-1)}{\mu(\mu-1)} \qquad (11)$$

where μ represents the rank of fittest individual and I denotes the rank of the current individual.

   Thus, from the present population, pair of fuzzy if then rules is chosen to produce new fuzzy rules for next generation.

4. Now a pair of fuzzy if then rule is randomly taken, and a crossover operation is applied.

   The cross over operation on the fuzzy if then rule is performed with specific crossover probability and is given as:-

Parent 1-If X_1 is l_1^1 and X_2 is l_2^3 and X_4 is l_4^2, then the class is C2 with $CD_{j1.}$

Parent 2-If X_1 is l_1^1 and X_2 is l_2^3 and X_4 is l_4^4, then the class is C2 with CD_{j2}

Offspring-If X_1 is l_1^1 and X_2 is l_2^3 and X_4 is l_4^2or l_4^4, then the class is C2 with $CD_{j3.}$

If the offspring generated give improved results and cover more samples, then it is accepted in the next generation otherwise it is rejected.

5. A feature is randomly chosen from the chromosome and bit flip mutation is applied to it with mutation probability. It means the value of the selected gene will be converted to 1 if its value corresponds to 0 or vice versa. This step is repeated until a specific number (M_R).

6. A local heuristic approach is applied on the gene pool where a chromosome is selected and value of the gene is randomly changed. If the obtained results are better than the previous one, then that rule is taken as an offspring otherwise it is rejected. Also the chromosomes with worst fitness function are removed to obtain good results. The process terminates when the specified number of generations are completed.

7. After performing all the above steps, a new rule set is obtained. Now a mathematical model is applied on it to obtain more accurate and optimized results.

Here are the following sets and parameters available on the basis of which more samples will be covered in the fuzzy if then rules and accuracy will be maximized as given by [21]:-

- C -represent set of classes
- F -set of features
- S -represent set of samples
- R -represent set of rules
- μ_{sr} -represent membership function of sample s for rule r.
- α_r -represent accuracy value of rule r
- C_s - denotes the class label for the sample s
- C_r - represent class label for rule r
- M -a very high number

Decision variables are given by:-

$$x_r = \begin{cases} 1 & \text{if rule r is selected} \\ 0 & \text{else} \end{cases}$$

$$y_{sc} = \begin{cases} 1 & \text{if sample s is classified as class c} \\ 0 & \text{else} \end{cases}$$

The model is given by:-

$$MaximizeT = \sum_{s \in S, c \in C, C_s = c} y_{sc} \tag{12}$$

It means that more number of samples should be classified correctly. Therefore, each sample must be classified into their correct class otherwise any attack can be misclassified as normal and may pose a serious threat to the network and this is done by maximizing the number of correct classification while satisfying the following conditions:-

$$\sum_{r \in R: C_r = c} x_r \geq 1, \forall c \in C \tag{13}$$

It represents that at least one rule is selected from each class.

$$\sum_{c \in C} y_{sc} \leq 1, \forall s \in S \tag{14}$$

It checks that each sample should be classified in maximum one class.

To obtain the preciseness of each selected rule, the accuracy is calculated as:-

$$Accuracy_s \geq \sum_{r \in R: C_r = c} \mu_{sr} \alpha_r x_r, \forall s \in S, \forall c \in C \tag{15}$$

where $Accuracy_s \geq 0 \; \forall s \in S$ is a constraint.

To ensure that for each rule, the corresponding class selected is correct, following formula is applied-

$$y_{sc} = 1 - (\frac{1}{M})(Accuracy_{sc} - \sum_{r \in R:C_r = c} \mu_{sr} \alpha_r x_r) \forall s \in S, \forall c \in C \quad (16)$$

Hence the main objective is to maximize T and thus cover approximately all the samples of training dataset. Therefore the feasibility of obtaining the best rule set will be maximized which will improve the classification of attacks in testing dataset as more features will be available for classification.

The overall steps have been given in the following flowchart (fig 3):-

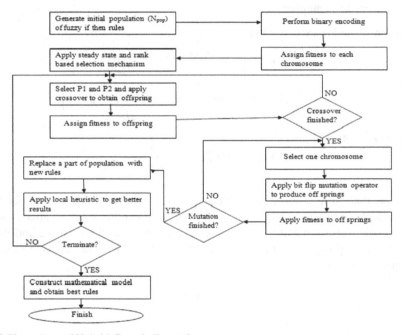

Fig. 3 Flow chart of Hybrid Genetic Fuzzy System

5   Experimental Setup and Results

The experiments have been laid down on KDD99 dataset to set up an intrusion detection system. This dataset contains standard audit data including a large number of intrusions. In the proposed approach, 10% of the KDD99 dataset has been used i.e. the total number of records are 494,021. There are two dataset available i.e. training and testing datasets. The training dataset is labeled specifying the behavior of the system.

The dataset contains a large number of connections and each connection is a sequence of TCP packets. Each connection consists of 41 attributes that are utilized in making rules. There are 5 classes in the dataset given as:-

Table 1 Classification of Attacks

Sno.	Classes	Subclasses
1	Normal	NA
2	Denial of Service(DOS)	back,land,neptune,pod,smurf,teardrop
3	User to Root(U2R)	buffer_overflow, loadmodule, multihop, perl, rootkit
4	Remote to User(R2L)	imap,phf,spy, ftp_write, guess_passwd, warezclient, warezmaster
5	Probe	Nmap, portsweep, satan, ipsweep

In the given experiment, 750 samples were randomly chosen from the training dataset and the value of each attribute was normalized in the unit interval [0,1]. The parameters which were used to make a genetic fuzzy rule based inference engine are given below:-

Table 2 Parameters needed for fuzzy inference engine

SNo	Parameter	Assigned value
1	Population Size(N_{pop})	30
2	Crossover probability(P_c)	0.9
3	Mutation probability($P_{m)}$	0.1
4	Mutation rate(M_R)	20
4	Maximum number of generations	100

The genetic fuzzy based intrusion detection system is formed using the above parameters and a pool of rule set is obtained which is checked with the testing dataset.

The rules are checked on 5,000 randomly selected testing connections and conclusions were drawn on the basis of following parameters:-

$$\Pr ecision = \frac{TP}{TP+FP} \quad (17)$$

$$\operatorname{Re} call = \frac{TP}{TP+FN} \quad (18)$$

where *TP* represents True Positive, *FP* is False Positive and *FN* denotes False Negative.

Here are the results of the proposed genetic fuzzy inference engine:-

Table 3 Results obtained from the proposed work

Classes	Recall (%)	Precision (%)
Normal	91.45	85.1
DOS	97.09	97.6
U2R	86.10	89.1
R2L	91.66	94.8
Probe	94.60	84.1

The recall or detection rate for each class comes out to be of admirable value especially for the u2r and r2l classes where different algorithms do not succeed in getting a high precision.

Different approaches such as Michigan, Pittsburgh and Iterative Rule Learning have also been previously applied on KDD-99 dataset to generate robust IDS. The results of these existing genetic fuzzy systems have been compared with the current approach and validated to check whether the consequences or the outcomes are more accurate and better than traditional approaches. Following is the graph which represents the consequences of different schemes on each class.

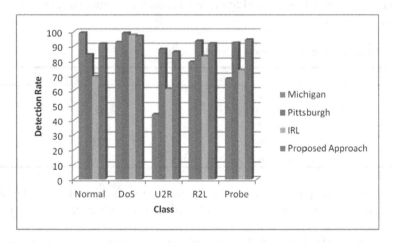

Fig. 4 Evaluation of Detection Rate on each class for different approaches

The Michigan approach performs well in identifying normal class as compared to all other approaches with high percentage of 99.2 but does not function as well in other classes. The proposed approach provides output of around 91% in normal connections. The Pittsburgh approach achieves high detection rate in case of U2R of around 87% and 93% in R2L. The Michigan and IRL approach performs poorly in detecting U2R class with recall of around 44% and 61%. Even in R2L the detection rate is at lower rate of around 79% in case of Michigan approach and 83.2 % in Iterative Rule Learning. The intended methodology gives nearly equivalent results in comparison to Pittsburgh approach with True Positive Rate as 86% in U2R and 92% in R2L. The analogous behaviour is shown by all the approaches in case of DoS attacks. The intended approach outperforms all the existing genetic fuzzy approaches with recall of 94.6% while Michigan and Iterative Rule Learning performs poorly with recall of 68.2% and 73.9%. The graph clearly depicts that the purported methodology outdoes Pittsburgh approach in case of detecting normal and probe class and give equivalent results in case of other classes too. Therefore, the proposed genetic fuzzy inference engine is a proficient approach which can be used to detect intrusions in network.

6   Conclusions

With the hybridization of genetic algorithm and a fuzzy rule based system, an efficient approach has been built to detect intrusions in an effective manner. The current approach is able to detect U2R and R2L attacks with high recall value which the other approaches fail to do so due to their less number in training dataset. Also with the establishment of mathematical model, best rules are preserved, thus providing a good approach to detect attacks. The classification rules are able to classify the normal and abnormal behavior in the network with good accuracy, thus leaving fewer loopholes for the misjudgment.

This approach can be made more effective by using better local search algorithms to generate more compact and accurate results. Also, the number of rules to be considered and kept in the database has not yet been evaluated. Thus the scope of this work can be extended by working on making concise and compact rules. Therefore, it would help in maximum classification in minimum number of rules.

References

1. Luger, G., Maccabe, A., Servilla, M.: The architecture of a network-level intrusion detection system. Department of Computer Science. College of Engineering, University of New Mexico (1990)
2. SANS Institute, Intrusion Detection Systems: Definition, Need and Challenges. SANS Institute (2001), http://www.sans.org/reading_room/whitepapers/detection/intrusion-detection-systems-definition-challenges_343
3. Ishibuchi, H., Yamamoto, T.: Fuzzy Rule Selection By Data Mining Criteria And Genetic Algorithms. In: GECCO, pp. 399–406 (July 2002)
4. Ishibuchi, H., Yamamoto, T.: Comparison of heuristic criteria for fuzzy rule selection in classification problems. Fuzzy Optimization and Decision Making 3(2), 119–139 (2004)
5. Nakashima, T., Ghosh, A.: Classification confidence of fuzzy rule-based classifiers
6. Denning, D.E.: An intrusion-detection model. IEEE Transactions on Software Engineering (2), 222–232 (1987)
7. Debar, H., Dacier, M., Wespi, A.: Towards a taxonomy of intrusion-detection systems. Computer Networks 31(8), 805–822 (1999)
8. Herrera, F., Magdalena, L.: Genetic fuzzy systems: A tutorial. Tatra Mt. Math. Publ. (Slovakia) 13, 93–121 (1997)

9. Middlemiss, M.J., Dick, G.: Weighted feature extraction using a genetic algorithm for intrusion detection. In: The 2003 Congress on Evolutionary Computation, CEC 2003, vol. 3, pp. 1669–1675. IEEE (December 2003)

10. Liao, Y., Vemuri, V.R.: Use of K-nearest neighbor classifier for intrusion detection. Computers & Security 21(5), 439–448 (2002)

11. Lee, C.H., Shin, S.W., Chung, J.W.: Network intrusion detection through genetic feature selection. In: Seventh ACIS International Conference on Software Engineering, Artificial Intelligence, Networking, and Parallel/Distributed Computing, SNPD 2006, pp. 109–114. IEEE (June 2006)

12. Wang, X., Yang, J., Teng, X., Xia, W., Jensen, R.: Feature selection based on rough sets and particle swarm optimization. Pattern Recognition Letters 28(4), 459–471 (2007)

13. Stein, G., Chen, B., Wu, A.S., Hua, K.A.: Decision tree classifier for network intrusion detection with GA-based feature selection. In: Proceedings of the 43rd Annual Southeast Regional Conference, vol. 2, pp. 136–141. ACM (March 2005)

14. Mukkamala, S., Sung, A.H.: Feature ranking and selection for intrusion detection systems using support vector machines. In: Proceedings of the Second Digital Forensic Research Workshop (August 2002)

15. Mukkamala, S., Janoski, G., Sung, A.: Intrusion detection using neural networks and support vector machines. In: Proceedings of the 2002 International Joint Conference on Neural Networks, IJCNN 2002, vol. 2, pp. 1702–1707. IEEE (2002)

16. Hofmann, A., Horeis, T., Sick, B.: Feature selection for intrusion detection: an evolutionary wrapper approach. In: Proceeding of the 2004 IEEE International Joint Conference on Neural Networks, vol. 2, pp. 1563–1568. IEEE (July 2004)

17. Lu, W., Traore, I.: A new evolutionary algorithm for determining the optimal number of clusters. In: 2005 and International Conference on Intelligent Agents, Web Technologies and Internet Commerce, International Conference on Computational Intelligence for Modeling, Control and Automation, vol. 1, pp. 648–653. IEEE (November 2005)

18. Gomez, J., Dasgupta, D.: Evolving fuzzy classifiers for intrusion detection. In: Proceedings of the 2002 IEEE Workshop on Information Assurance, vol. 6(3), pp. 321–323. IEEE Computer Press, New York (2002)

19. Abadeh, M.S., Mohamadi, H., Habibi, J.: Design and analysis of genetic fuzzy systems for intrusion detection in computer networks. Expert Systems with Applications 38(6), 7067–7075 (2011)

20. Berlanga, F.J., Rivera, A.J., del Jesús, M.J., Herrera, F.: GP-COACH: Genetic Programming-based learning of COmpact and ACcurate fuzzy rule-based classification systems for High-dimensional problems. Information Sciences 180(8), 1183–1200 (2010)

21. Aydogan, E.K., Karaoglan, I., Pardalos, P.M.: hGA: Hybrid genetic algorithm in fuzzy rule-based classification systems for high-dimensional problems. Applied Soft Computing 12(2), 800–806 (2012)

Enhancement of Data Level Security in MongoDB

Shiju Sathyadevan, Nandini Muraleedharan, and Sreeranga P. Rajan

Abstract. Recent developments in information and web technologies have resulted in huge data outburst. This has posed challenging demands in efficiently storing and managing large volume of structured and unstructured data. Traditional relational models exposed its weakness so much so that need for new data storage and management techniques became highly desirable. This resulted in the birth of NoSQL databases. Several business houses that churn out large volume of data have been successfully using NoSQL databases to store bulk of their data. Since the prime objective of such DB's were efficient data storage and retrieval, core security features like data security techniques, proper authentication mechanisms etc. were given least priority. MongoDB is one among the most popular NoSQL databases. It is a document oriented NoSQL database which helps in empowering business to be more agile and scalable. As MongoDB is gaining more popularity in the IT market, more and more sensitive information is being stored in it and so security issues are becoming a major concern. It does not guarantee privacy of information stored in it. This paper is about enabling security features in MongoDB for safe storage of sensitive information through "MongoKAuth" Driver, a new MongoDB client side component developed in order to automate a lot of manual configuration steps.

Keywords: NoSQL, MongoDB, Security, Kerberos.

Shiju Sathyadevan
Amrita Cyber Security Center,
Amrita Vishwa Vidyapeetham, Amritapuri, India
e-mail: shiju.s@am.amrita.edu

Nandini Muraleedharan
Department of Cyber Security Systems and Networks,
Amrita Vishwa Vidyapeetham, Amritapuri, India
e-mail: neelima.nandini@gmail.com

Sreeranga P. Rajan
Fujitsu Laboratories, US
e-mail: sree.rajan@us.fujitsu.com

© Springer International Publishing Switzerland 2015   199
R. Buyya and S.M. Thampi (eds.), *Intelligent Distributed Computing*,
Advances in Intelligent Systems and Computing 321, DOI: 10.1007/978-3-319-11227-5_18

1   Introduction

The various developments in information technology and web technology are demanding the storage of large amount of unstructured data in distributed databases which in turn should provide high availability and scalability. Many companies are now using NoSQL databases to store bulk of their data. These are non-relational databases without full SQL functionality. Usually NoSQL databases have simple and flexible data model and are highly scalable and reliable. But they do not have proper mechanism to maintain the integrity and confidentiality of data. MongoDB is the most popular NoSQL databases widely used in the IT world. It helps in empowering business to be more agile and scalable.

MongoDB is a document oriented database written in C++ language and handles collections of JSON like documents [1]. It encodes data in BSON format [1]. It is free and open source software. In contrast to traditional table based relational database, it does not have any predefined schemas. This feature of mongodb makes the integration of data in certain types of applications easier and faster. It supports indexing, full query language and consistency. The Mongo database was developed by 10gen (now MongoDB Inc.) for scalability, performance and high availability, scaling from single server deployments to multi cluster architectures [2]. As MongoDB is now gaining more popularity in the IT market more and more sensitive information are being stored in it and security issues are becoming a major concern. It does not guarantee the protection and privacy of information stored in it. There are many inbuilt security issues within this database. This database is built in such a way that by default there is no authentication and authorization of users enabled in it. It is because of the assumption that the database will be running only in a secure environment. The data at rest is unencrypted so that it will always be made available to the users accessing it. By default there is no proper auditing in MongoDB. This database is also prone to injection attacks as it heavily utilizes JavaScript as an internal scripting language.

The addition of security features in MongoDB is very much a user requirement, that it depends on the mongo users to enable or remove the available security. The security need to be enabled in such a way that it should not affect the performance of the database. In this paper we discuss about how to improve the security features in MongoDB for safe storage of sensitive information. We start with showing a threat model in section 3. We then discuss about an easy to configure client side Kerberos authentication driver for MongoDB in section 4, performance analysis in section 5 and finally conclude in section 6.

2   Related Works

Most of the NoSql databases such as Cassandra and MongoDB have similar security issues. Both of them do not have proper authentication and authorization mechanism to ensure the integrity and confidentiality of information stored in them. It is the administrator who has to deal with these kinds of security problems. The communication between these databases and clients are not encrypted [1]. An intruder monitoring the data traffic can easily find the transmitted data through the network in plain text format. Also these databases do not have proper auditing enabled in them.

Big data systems are those which can store large amounts of data, handle data across many systems and provide facilities for data queries, data consistency, and systems management [3]. These systems mostly make use of one or more Hadoop components and then extend some of the basic functionality [3]. The Security recommendations for big data systems include use of Kerberos; File/OS layer encryption, Key management, and node validation during deployment, log transactions, anomalies and administrative activity, use of SSL/TLS network security etc. These techniques provide protection against normal attacks with minimum hindrance to applications and operations. The security issues in the big data environment are Architectural security issues and Operational security issues. Architectural issues include distributed nodes, sharded data, data access/ownership, issues in inter node communication and client interactions [3]. These security issues are not inherent to big data, but big data projects are inherently subject to them because of the distributed architecture, use of simple programming model and the open framework of services [3]. Operational security issues arise in different context of database management systems and they are: Data at rest protection, administrative data access, configuration and patch management, authentication of applications and nodes, auditing and logging problems, API security etc. The security measures used for big data must not compromise the basic functionality of a cluster and it's characteristics. It should also address the security threat to big data environments.

Most of the NoSQL databases have been developed for bulk storage and high availability of data and so they often lag behind in incorporating proper security features to protect the data stored in them. But it doesn't mean that we cannot enable security in them. We can make use of third party tools to ensure security of data. One such solution is IBM's InfoSphere Guardium Data Encryption. It sits between the OS file system and the database. An important feature of this tool is that it is completely transparent to the database and the applications and also it can be used in heterogeneous environments to encrypt structured and unstructured data. There are policies developed which provide the ability to specify which users (processes) can decrypt the data, and even which files/directories should be encrypted etc. This also helps in monitoring database activities and protecting data in motion. InfoSphere Guardium continuously monitors data activity using lightweight software probes, called S-TAPs, without relying on logs. The S-TAPs also do not require any configuration changes to the NoSQL servers or applications [5]. There are three major security vulnerabilities that should be addressed by application developers that use NoSQL data store i.e., insecure connection between web application and database, insufficient support for special authorized users (e.g., DBA) and insufficient authentication [6]. Many NoSQL databases provide an interface that serializes objects directly into NoSQL storage (e.g. documents); if this interface is exposed directly to the user it can be abused by employing injection attacks. The best method is to place the security at the middleware level. Now let us consolidate our discussion about enhancing security features to MongoDB alone and throw light on some of the security issues in it.

3   Threat Model

Based on the different security weaknesses in MongoDB a threat model is developed for different security levels. The different security levels focussed are: Document level, Client level and Database level.

3.1   Document Level

MongoDB does not assure security of data files stored in it. The data files are not encrypted; as a result anyone can extract information from the data files. There is no mechanism available to encrypt it. We can mitigate this weakness by employing file system level encryption and by setting access permissions.

3.2   Client Level

In MongoDB there is no security to the client communicating with the database. Wire level protocols are used by most of the drivers for communication with database and for replication. Since wire protocols are unencrypted and uncompressed, it is not secure. An attacker can monitor network traffic because there is no encryption of information. To overcome this vulnerability packet filter rules are added to prevent unknown host connections.

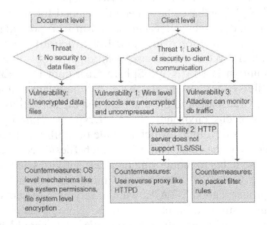

Fig. 1 Diagram showing document level and client level threat models

3.3   Database Level

MongoDB is subjected to Injection attacks, where un-trusted data is send to an interpreter as part of a command or query. Such data can trick the interpreter to execute unintended commands or access unauthorized data. In this the internal commands available to the developers are short scripts written using JavaScript.

JavaScript is an interpreted language, as a result an attacker can alter query statements using "where" clause. To overcome this injection attacks, proper sanitization of user inputs is necessary [1]. Inter-node or Inter-cluster communication is not secure since nodes on a cluster can freely communicate without any encryption and authentication. The database does not have any strong authentication and authorization mechanisms by default. It does not ensure privacy of communications between nodes. There is no auditing existing in this database and no data logs available to record the events taking place in the database to understand and diagnose security problems. Kerberos can be used to validate nodes and client applications before admission to the cluster. Third party logging tools are available to get the data logs to diagnose any problems but are rarely used.

Relational databases like MySql, Oracle etc are shipped with auditing tools that allows administrators to monitor the database to maintain security of information. In case of Mongodb, every time when a new user is added to the mongo database, their details are logged. But later on throughout the life of the database they are not audited. Due to this the activities of the users cannot be monitored and any intruder who impersonate as a database user can perform illegal activities. While taking into account different security threats at these levels of MongoDB, one can come to a conclusion that there is no security of any kind to protect sensitive data stored in MongoDB.

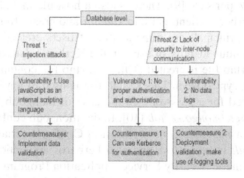

Fig. 2 Diagram showing database level threat model

4   Enabling Kerberos Authentication in MongoDB

In MongoDB Kerberos can be enabled either manually or programmatically. There are several vital configuration steps involved in this; missing out any of the vital steps can result in unsuccessful attempt to configure Kerberos. This paper mainly discusses about how Kerberos can be configured programmatically.

4.1 Kerberos Authentication

We propose to employ Kerberos for user authentication in MongoDB as it is a strong authentication mechanism based on cryptography. It allows the transmission of password entered by a user during login in encrypted form through the network. This prevents any attacker from intercepting the user password and misusing it to do unauthorized activities. It makes use of symmetric key cryptography for encrypting the user passwords. It also supports mutual authentication, where both the user and the server verify each other's identity. It provides protection against replay attacks and eavesdropping. In recent years different versions of MongoDB have been released by 10gen. Among them Kerberos authentication is supported only in Enterprise edition of MongoDB.

4.2 Kerberos Manual Configuration

There are several requirements to be satisfied for building a strong authentication mechanism in MongoDB. There must be a Kerberos client and a Kerberos server configured for this purpose. As an initial step, an enterprise edition of MongoDB is installed and a Kerberos client is configured for it. The client requests the Kerberos credential from the key distribution centre (KDC) on behalf of mongo user. A Kerberos client consist of two configuration files namely *krb5.conf* and *kdc.conf* as shown in Config.2 and Config.3. The *krb5.conf* file contains details about the Kerberos realm, the admin server, and the KDC [9]. The *kdc.conf* file contains the details about the Kerberos database path, admin keytab path, the encryption schemes used, *kdc* ports etc [9]. There is also a host file included with it. After the installation of Kerberos client, get into *kadmin* and create Kerberos principals for mongodb. These principals are same as that of users created in the mongo database. These are unique identity of users in the Kerberos database. For each principal created, a keytab file is created [10]. A keytab file contains pairs of Kerberos principals and encrypted keys. This keytab file is used to log in to Kerberos without being prompted for a password. In the case of mongodb, the keytab file created is moved to */etc/mongo.keytab* [10]. In the mongodb configuration file named as *mongod.co*nf or *mongodb.conf* as shown in Config.1, certain modifications are made such as *auth* is set to *true* and also kerberos authentication mechanism is specified in it. Generic Security Service Application Program Interface (GSSAPI) is the authentication mechanism employed here [10]. It is a collection of functions that provides security services. This application program interface is used by clients and servers to authenticate to each other without either program having specific knowledge about the underlying mechanism. Whenever any network service uses GSS-API, it can authenticate using Kerberos.

```
#/etc/mongod.conf,Example configuration file.

fork = true
auth = true
dbpath = /opt/mongodb/data
logpath = /opt/mongodb/log/mongod.log
setParameter = authenticationMechanisms=GSSAPI, MONGODB-CR
```

Config. 1 MongoDB configuration file

KDC issues the tickets to the clients to authenticate with the mongo database [9]. There are two types of tickets issued to the Kerberos principal. One is the Ticket Granting Ticket (TGT) and the other is the service ticket [9]. The TGT is issued by KDC to the client to request the service ticket from the Ticket Granting Server (TGS). The TGS verify the validity of the ticket and issues a service ticket and a session key to the client to access the required service from the mongo server. At the end in order to make the authentication of users successful, we need to invoke the Kerberos along with the keytab file and the security mechanism within the server manually.

```
kdc_ports = 750,88

[realms]
    DATA.ACL.NET = {
        database_name = /var/lib/krb5kdc/principal
        admin_keytab = FILE:/etc/krb5kdc/kadm5.keytab
        acl_file = /etc/krb5kdc/kadm5.acl
        key_stash_file = /etc/krb5kdc/stash
        kdc_ports = 750,88
        max_life = 10h 0m 0s
        max_renewable_life = 7d 0h 0m 0s
        master_key_type = des3-hmac-sha1
        supported_enctypes    =    arcfour-hmac:normaldes3-hmac-sha1:normaldes-cbc-
crc:normal des:normal des:v4 des:norealm des:onlyrealm des:afs3
        default_principal_flags = +preauth
    }
```

Config. 2 Kerberos kdc.conf file

```
[libdefaults]
        default_realm = MYREALM.ME1
        forwardable = true
        proxiable = true

[realms]
        MYREALM.ME1 = {
                kdc = amcskdc.myrealm.me1
                admin_server = amcskdc.myrealm.me1
        }

[domain_realm]
        .myrealm.me1 = MYREALM.ME1
        Myrealm.me1 = MYRAELM.ME1

[login]
#       krb4_convert = true
#       krb4_get_tickets = false

        kdc = FILE:/var/log/krb5/kdc.log
        admin_server=FILE:/var/log/krb5/kadmin.log
default = FILE:/var/log/krb5/kadmin.log
```

Config. 3 Kerberos krb5.conf file

4.3   Issues In Manual Configuration

During the process of configuring Kerberos in MongoDB, we had to deal with several issues.

1.   Initially, a connection between the Kerberos client and the server was not successfully established due to the misconfiguration in the installation of Kerberos client. The Kerberos server name might not be correctly specified in the krb5.conf file (as shown in Config.3). Due to this client cannot get the valid Kerberos credential for its mongo user.

2.   Secondly there are times when the keytab files created for a Kerberos principal may not function properly. This happens when Kerberos invocation is not done (as shown in Config.4) efficiently. Also the Kerberos ticket generated for the user gets expired. For every ticket (TGT) generated, there will be an expiry date specified in it, up to which the ticket will be valid to carry out authentication process. But as soon as the ticket expires, it needs to be renewed to get the required credentials for successful authentication.

3.   Thirdly a user present in the mongo database may not be authenticated using Kerberos due to lack of permission. This happens when *auth* is set as *false* in *mongod.conf* file shown in Config.1. Due to this users are not authenticated before gaining access to the mongo database. There are several such parameters that should be set in various configuration files for successful deployment of Kerberos.

4.   The fourth concern is that the GSSAPI credential may not be obtained properly for authentication during the Kerberos invocation. This is because by default the *GSSAPI* security mechanism is not specified in the *mongod.conf* file.

5.   In manual configuration we need to invoke the Kerberos [10]. It is shown in Config.4.

```
env KRB5_KTNAME=/opt/mongodb/mongod.keytab \
/opt/mongodb/bin/mongod --config /opt/mongodb/mongod.conf
```

Config.4 Kerberos invocation manually

4.4   Kerberos Implementation Automation Using MongoKAuth Driver

MongoDB supports both normal authentication of users using native username/password and also Kerberos authentication [11]. Kerberos authentication is recommended in order to have stronger authentication. Authentication using password is subjected to brute force attacks and is not secure. User authentication using password is shown in Snippet 1.

```
# MongoCredential credential1 = MongoCredential.createMongoCredential (" stud1 ", "
cyber ", " passwd1 ". toCharArray() );

#MongoClient  mongoClient1 = new  MongoClient  (new  ServerAddress  (server),
Arrays.asList (credential1));
```

MongoKAuth Snippet. 1 Authentication using password

Here "stud1" is the user, "cyber" is the database name and "passwd1" is the password. Authentication using Kerberos can be done as shown in Snippet.2.

```
#MongoCredential     credential1     =     MongoCredential.createGSSAPICredential
("stud1@MYREALM.ME ");

#MongoClient    mongoClient1 = new  MongoClient  (new  ServerAddress(server),
Arrays.asList  (credential1) );
```

MongoKAuth Snippet. 2 Authentication using Kerberos

Here "stud1@REALM.ME" is the Kerberos principal of the user present in the database. GSSAPI is the Kerberos authentication mechanism applied here.

In order to make the Kerberos configuration easier, a new driver called "MongoKAuth" driver is implemented which is integrated as part of Mongo client. Here Kerberos is configured at client side using MongoKAuth driver and not at the mongo server side. It acts as a client which request services from mongo server on behalf of mongo users as shown in Fig 3. In the client program we need to specify the Kerberos principal name and the credentials of the user who need to be authenticated using Kerberos. Through this only after proper authentication, users can execute database queries concurrently and retrieve information.

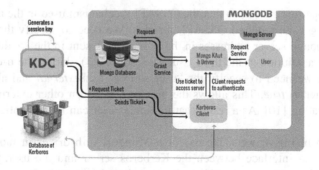

Fig. 3 Security architecture of MongoDB

In manual configuration of Kerberos, it is a tedious task for the administrator to authenticate multiple users at the same time. Tickets have to be renewed repeatedly every time for all the users before authentication, but in this approach tickets will be periodically renewed based on the system properties set within the client program. One does not want to invoke Kerberos and the keytab files repeatedly before authenticating each and every user. During the execution of the program the service ticket for the users are generated with which users can easily access the database. Everything happens automatically soon after the execution of the program. Since all these are done at the client side, server side overhead can be reduced and query processing time is not much affected. This approach is more efficient as it reduces the burden of the administrator and it is very simple from a user's point of view as the configuration steps involved is very less. Several issues raised during the manual configuration can be easily solved via this programmatic configuration of Kerberos using MongoKAuth driver. The first issue in the manual configuration (mentioned in Section 4.3.1) can be solved by correctly identifying the Kerberos realm and the admin server for the Kerberos client and dynamically configuring it within MongoKAuth Driver. The second, fourth and the fifth issues (Section 4.3.2, Section 4.3.4 & Section 4.3.5) can be corrected by setting certain system properties within the driver [11]. The setting of system properties is shown in Snippet.3.

```
#System.setProperty("java.security.krb5.conf",  " /etc/krb5.conf ");

#System.setProperty("java.security.krb5.realm",  " MYREALM.ME1 ");

#System.setProperty ("java.security.krb5.kdc", "amcskdc.myrealm.me1 ");

#System.setProperty("javax.security.auth.useSubjectCredsOnly",  " false ");

#System.setProperty("sun.security.krb5.debug",  "true");
```

MongoKAuth snippet.3 System properties

These properties are set, so that the GSSAPI mechanism used in the authentication process get the required GSSAPI libraries. We need to specify the Kerberos server and mongo server addresses in the *hosts* file present in the *etc* directory. In order to find a solution to the third issue (Section 4.3.3), during the user creation in mongodb, we need to specify the "user source" as *$external* and also set the *auth* parameter to *true*. This allows the user to depend on other external sources for authentication [10]. As a result that particular user can be authenticated using Kerberos.

As shown in Fig 4, we make use of a MongoKAuth driver in mongo client which act as an interface between the Kerberos server and the user. Instead of simply providing services to different users directly, it authenticates each and every user accessing the mongo database using Kerberos.

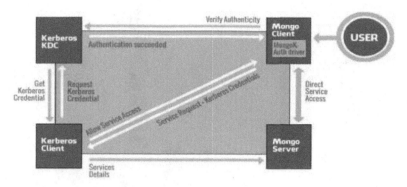

Fig. 4 Kerberos authentication mechanism flow diagram

We need to set up a Kerberos client and create principals and keytab files for each and every user which has to be authenticated via Kerberos. There are certain inbuilt functions for defining Kerberos client and Kerberos principal in the client program. The authentication mechanism, GSSAPI will be specified in the program. Mongod, an internal mongodb process, should be up and running during the process of authentication. The mongo database must have users in the name of different Kerberos princi-pals present in the Kerberos database. The authentication method for each user in the database must be specified as *$external*. After integrating the new driver one can see the internal configuration processes taking place during authentication without any manual intervention whereby service tickets are granted automatically for each MogoDB user to access the required data. Whenever a user wants to access mongo data-base, a MongoKAuth driver client requests KDC for that particular user's Kerberos credential. If valid credential is present, then MongoKAuth driver issues the same to the Kerberos client. With the obtained credential, mongo client requests service from the mongo server. Then KDC verifies the authenticity of the mongo client and authentication is successfully done. After this mongo server allows service access to the user. In this way user request is satisfied and the required service is granted to the user.

5   Performance Analysis

For analysis mainly two scenarios are considered, one is time taken for the upload and retrieval of files of different size with and without Kerberos authentication. The second scenario is the time taken to access different Mongo database instances (DBi) by a user with and without Kerberos authentication. Fig 5 depicts the first performance analysis scenario.

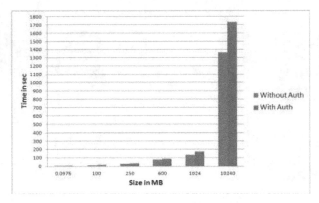

Fig. 5 Graph showing the time taken (in seconds) for uploading and retrieving files of different size

Here we tried to upload files of different sizes ranging from 100MB to 1GB onto MongoDB with and without enabling Kerberos authentication. We found that depend-ing on the size of the files i.e., when file size is larger and authentication persist; it is consuming slightly more time to handle the files as compared to the scenario where there is no Kerberos authentication enabled. Though the time con-sumption is slightly more, it is providing more security to the data files in the database.

According to the second scenario, we have considered two types of authentica-tion on a particular Mongo user. In Kerberos authentication, the user will be pro-vided with a service ticket which will be enabled for the entire session, while for password au-thentication, the user need to provide the password for each access to a particular Mongo Database instance (DBi). Even though service ticket granting for a session during its very first phase is time consuming, in subsequent sessions this is cached and hence the same service ticket can be used for each access of the same session. From this, it is evident that the Kerberos provides much stronger authentication without

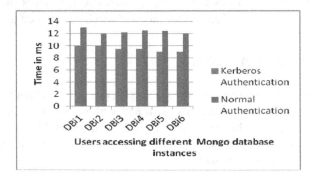

Fig. 6 Graph showing the time taken in milliseconds for a user to query on different Mongo database instances and retrieve data when authenticated using Kerberos and when authenti-cated using password

heavy impact on performance (as shown in Fig 6) compared with the native Mon-goDB password authentication. In a broad sense, when Kerberos user is accessing different DBis', although initial query processing takes a bit longer time, further query processing will be executed with no delay. Another privilege provided to the Kerberos user is the session independent TGT generated in the initial phase. The use of the cached copy of TGT, which can be renewed on expiry, makes Ker-beros authentication time efficient.

6   Conclusion

In this paper an attempt is being made to throw light on different security weak-nesses in MongoDB, which in general is applicable across all NoSQL databases. Paper also details the development and integration of a new MongoDB client side driver that can automate Kerberos authentication, thereby easing the manual deployment process in mitigating one of the major threat i.e., lack of proper au-thentication. To protect the sensitive information stored in MongoDB, and also to maintain confidentiality, stronger authentication mechanism is employed by using Kerberos. This is now integrated with "Amrita Big Data FrameWork" (ABDF) which is a major open source all integrated big data processing research initiative which when completed will have the capability to process data either in linear fashion, or using single node or cluster based hadoop framework or using in-memory analytic engines like SPARK/STORM or with in-database analytics, all of these integrated under one framework using an easy to use process flow defini-tion mechanism.

References

1. Okman, L., Gal-Oz, N., Gonen, Y., Gudes, E., Abramo, J.: Security issues in NoSQL databases,
   http://jmiller.uaa.alaska.edu/csce465fall2013/papers/
   okman2011.pdf
2. MongoDB overview, http://www.mongodb.com/mongodb.overview
3. Lane, A.: Securing big data-Security recommendations for hadoop and NoSql envi-ronment, https://securosis.com/assets/library/reports/
   SecuringBigData_FINAL.pdf
4. Bhatewara, A., Waghmare, K.: Improving network scalability using NoSQL database. IJACR (December 2012)
5. NoSQL does not have to mean no security,
   http://public.dhe.ibm.com/common/ssi/ecm/en/nib03019usen/
   NIB03019USEN.PDF
6. Virtual panel: Security consideration in accessing NoSQL databases,
   http://www.infoq.com/articles/nosql-data-security-
   virtual-panel
7. Kaur, H., Kaur, J., Kaur, K.: A review on non-relational databases, their types, advan-tages and disadvantages. IJERT (February 2013)

S. Sathyadevan, N. Muraleedharan, and S.P. Rajan

8. MongoDB-Security weaknesses in a typical NoSQL database,
   http://blog.spiderlabs.com/2013/03/mongodb-security-
   weaknesses-in-a-typical-nosql-database.html
 9. Kerberos ubuntu documentation,
   https://help.ubuntu.com/community/kerberos
10. Deploy mongodb with kerberos support,
   http://docs.mongodb.org/manual/tutorial/
   control-access-mongodb-with-kerberos-authentication
11. Authenticate to mongodb with java driver,
   http://docs.monggodb.org/ecosystem/tutorial/
   authenticate-with-java-driver/
12. Kerberos- community help wiki,
   https://help.ubuntu.com/community/kerberos
13. MongoDB, http://en.wikipedia.org/wiki/MongoDB
14. Lakshman, A., Malik, P.: Cassandra: a decentralized structured storage system. SIGOPS Oper. Syst. Rev. 44, 35–40 (2010),
   http://doi.acm.org/10.1145/1773912.1773927
15. Securosis blog-NoSQL and no security,
   http://securosis.com/blog/nosql-and-nosecurity
16. MongoDB, Officialwebsite, http://www.mongodb.org/
17. NoSql does not have to mean no security, http://IBM-public.dhe.ibm.com/common/ssi/ecm/en/../NIB03019USEN.PDF
18. Neuman, B.C.: Kerberos: an authentication service for computer networks. Inf. Sci. Inst., Univ. of Southern California, Marina del Rey, CA, USA
19. Kerberos, http://www.centos.org/docs/5/html/Deployment_Guide-en-US/chkerberos.html
20. Arora, R.P.: Head of the Department, Computer Sc and Engg; Dehradun Institute of Technology, Garima Verma; Implementation of authentication and transaction security base on Kerberos. IJITCE (February 2011)

SemCrawl: Framework for Crawling Ontology Annotated Web Documents for Intelligent Information Retrieval

Vandana Dhingra and Komal Kumar Bhatia

Abstract. Web is considered as the largest information pool and search engine, a tool for extracting information from web, but due to unorganized structure of the web it is getting difficult to use search engine tool for finding relevant information from the web. Future search engine tools will not be based merely on keyword search, whereas they will be able to interpret the meaning of the web contents to produce relevant results. Design of such tools requires extracting information from the contents which supports logic and inferential capability. This paper discusses the conceptual differences between the traditional web and semantic web, specifying the need for crawling semantic web documents. In this paper a framework is proposed for crawling the ontologies/semantic web documents. The proposed framework is implemented and validated on different collection of web pages. This system has features of extracting heterogeneous documents from the web, filtering the ontology annotated web pages and extracting triples from them which supports better inferential capability.

Keywords: Semantic Web, Ontologies, Crawling, Resource Description Framework (RDF), DARPA Agent Markup Language (DAML), Web Ontology Language (OWL), Uniform Resource Identifier (URI).

Vandana Dhingra
Department of Technology,
University of Pune, Pune
e-mail: vandana_dua_2000@yahoo.com

Komal Kumar Bhatia
YMCA University of Science & Technology,
Faridabad
e-mail: komal_bhatia1@rediffmail.com

© Springer International Publishing Switzerland 2015   213
R. Buyya and S.M. Thampi (eds.), *Intelligent Distributed Computing*,
Advances in Intelligent Systems and Computing 321, DOI: 10.1007/978-3-319-11227-5_19

1 Introduction

The World Wide Web (WWW) has revolutionized the means of data availability on the internet [6]. With the current structural model of the World Wide Web where anyone can easily publish its own document leads to lot of unstructured information and abundance volume. This is posing difficulties for current web crawlers and search engines to gather relevant information and henceforth it is becoming difficult for users to find information with proper precision and recall.

These difficulties have emerged with a solution to the extension of current web termed as semantic web. The semantic web is an extension of the current web in which information is given well defined meaning, better enabling computers and people to work in co-operation [1]. In particular, the semantic web provides a mechanism that is very useful for formatting data in machine readable form, linking individual data properties to globally accessible schemas, matching local references to entities against various kinds of standard names, and providing a range of inferences over that data in scalable ways [12]. The main components that distinguishes Semantic Web are –Ontologies, Languages used to represent ontologies, schemas to represent concepts, adding meaning to document data, triples, URI [5]. It is found that semantic markup within documents leads to greater number of relevant documents as compared with text documents [13].

In present there are different kinds of resources in semantic web – HTML documents embedded with metadata, RDF, OWL,DAML[4], RDF embedded XML documents, all these resources cannot be crawled by the current crawlers because of the meaning based linkage as compared to keyword based linkage in traditional web. Hence there is need of design of system that can crawl all these resources, extract the information from the semantic annotated documents which in turn will help out in providing inferential capability. The above discussed requirement for crawler is designed and implemented in this paper.

The paper is organized as: section 2 discuss about the motivation for developing the crawler framework; section 3 describes the related work in this area; section 4 discusses the proposed crawler and the corresponding algorithms for crawling semantic web documents and ontologies; in section 5 implementation of the crawler framework is presented with the validation of the work done and finally in section 6 conclusion is given.

2 Motivation

Crawling the semantic web is different from crawling the web of HTML documents. A traditional crawler starts with some seed URLs, downloads the corresponding documents, analyzes each document to gather further URLs for crawling and does context specific processing of the retrieved contents, like creating the searchable entries in the database. The last steps are repeated until a stop criterion is met (e.g. no more URLs to crawl, reached a predefined link depth, or gathered a predefined amount of documents) [14]. But these steps cannot be applied for crawling the web documents specified in languages described for the semantic web.

Table 1 Comparison between the traditional web and semantic web

Parameter	Traditional Web	Semantic Web
Basic concept	It is a collection of documents linked by hyperlinks described in languages with syntax that involves keywords.	It is a collection of documents linked by relations with inferential capability, hence adding a meaning to the links as compared to linkages merely via keywords.
Linkages Structure	Linked using an HTML anchor tag (link) which is a keyword reference to another document generally displayed as underlined text.	Linking among the documents are implemented using the rdfs:seeAlso[17] relationship[11]
Linkage Specification	An HTML hyperlink doesn't specify the actual meaning based linkage between documents.	Documents are written in Resource Description Framework (RDF), which make it possible to specify how concepts are linked to each other.
Crawling	Operates on HTML documents.	Semantic Web crawler operates on RDF, OWL and other semantic web representation languages [10].

Table 1 specifies the difference between the traditional web and semantic web crawler hence arising the need for design of crawler with different features as compared to the crawlers for crawling the traditional web documents. The functioning of semantic crawler is differentiated from normal crawler that normal crawler must only contend with extracting text from possibly invalid, content and subsequent link extraction whereas a semantic web crawler must carry out additional processing task like merging of information resources via inverse-functional- properties; tracking provenance of data; harvesting schemas and ontologies in addition to source data [16]. The above tabular representation clearly specifies that there is a difference between the linkage structures and linkage specification between the two web structures. Because of these different structures regular crawlers are not sufficient to crawl semantic web and special semantic web crawlers should be developed [15]. Hence there is requirement for crawling framework with different specialties to harvest the semantic web and create knowledge base.

3   Related Work

Ontotext RDF crawler [16] downloads interconnected fragments of RDF from the World Wide Web and builds a knowledge base from this data. At every phase of RDF Crawling, a list of URIs to be retrieved as well as URI filtering conditions (e.g. depth, URI syntax) are maintained which is done to download the resources containing RDF iteratively. To enable embedding in other tools, RDF Crawler provides a high-level programmable interface (Java API).

"Slug" a web crawler (or "Scutter") [7] is designed for harvesting semantic web content. Implemented in java using the Jena API, it works like web crawler, but it fetches RDF files instead of HTML pages, and follows rdfs: seeAlso links instead of HTML links. It has been designed as command based crawler system and does not provide any methods to reuse the crawled data and hence limiting the scope to just crawling of few documents and not providing the crawled data reuse. A Semantic crawler based on extended CBR algorithm [9] refers to harvesting semantic web contents by crawler which abstract metadata from online web pages and cluster them by associating with ontological concepts which is based on CBR algorithm. Swoogle crawler [8] can harvest, parse and analyze semantic web documents or semantic web information pieces embedded in web documents. RDF crawler [2] which is a multithreaded implementation capable of downloading simultaneously from many sources while aggregation thread does the processing. It builds a model that remembers the provenance of RDF and takes care to delete and replace triples if it hits the same URL twice.

Based on the above literature it is found very limited work is done in semantic web crawlers that focuses on extracting information from ontology annotated documents that will help in better information retrieval. The proposed work is different from all these related work, as it is focused on crawling the semantic web documents annotated with ontology for harvesting the knowledge base to produce more inferential results. A novel framework is developed that works on ontology annotated web pages and extract the triples relation from the underlying ontologies associated with the web page.

4   Proposed Framework

SemCrawler is the proposed crawler having features that crawls the HTML pages annotated with RDF/OWL ontology. The proposed crawler incorporates a filter module that filters out the HTML web pages and crawls the documents annotated with ontologies. The proposed architecture is shown in Fig.1. The proposed framework consists of following functional components:

    i. Fetch Module
    ii. Filter Module
    iii. Link Extraction Module
    iv. URI Dispatcher Module
    v. Parser Module

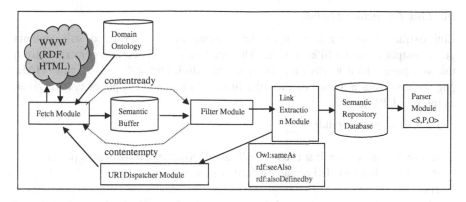

Fig. 1 Architecture of Proposed SemCrawl Framework

i. Fetch Module

It is a module that fetches HTML, RDF contents, corresponding ontologies associated with the web. After fetching, it stores, the web contents and ontologies in a semantic buffer. When the contents are transferred by fetch module to semantic buffer, signal "contentready" is sent to the filter module that filters the HTML web contents. On getting the "contentempty" signal from filter module, this module starts fetching the documents again from the World Wide Web.

Fetch Module
{
      wait (contentempty);
      extract URI from the URI queue;
      fetch the contents from web;
      store the contents in semantic buffer;
      signal (contentready);
}

ii. Filter Module

Filter module waits for the signal from the fetch module. After receiving the signal from fetch module, it gets the fetched contents stored in semantic buffer, filters the HTML web contents and the filtered contents are given as input to link extraction module for further processing sending the "contentempty" signal to fetch module for further fetching of contents from the web.

Filter Module (input: semantic buffer, output: filtered pages)
{
      wait (contentready);
      input the contents from buffer;
      extract the rdf web pages;
      input the filtered contents to link extraction module;
      signal (contentempty);
}

iii. Link Extraction Module

Link extraction module looks upon the contents which the filtered web contents got as output from the filter module. This module extracts certain constructs from the web page which works same as hyperlink <link href = "   ">in HTML page representation, hence extracts the further links from the pages is then given as input to URI queue.

Link Extraction Module
{
if (Owl: sameAs or rdf: seeAlso or rdf: alsoDefinedby constructs in page)
input those links to URI queue for further processing;
else
  store the web page in semantic repository database for further extraction of concepts;
}

iv. URI Dispatcher Module

This module gets the input from link extraction module, a list of URIs, which are given as input to fetch module for further downloading of semantic web documents from the web.

v. Parser Module

This module gets the input from semantic repositories database and will parse the web contents.

Parser Module
{
input the crawled data from semantic repository;
extract subject, predicate and object from the crawler output repositories;
store triples in database with three columns subject, predicate, object<S, P, O>;
}

5   Experiments and Results

For the purpose of first experiment 20 web pages were taken, out of which 5 web pages were HTML web pages and rest 15 pages were associated with ontology related to laptop domain and then gradually the number of pages were increased for subsequent tests as shown in Table 2. SemCrawl crawler was implemented in java using eclipse framework. Crawler was able to crawl all the web pages and ontology with filtering out the HTML web pages. This ontology annotation will help in finding relation between entities of the web page which will increase the further scope of research that the domain ontology developed by this research could be extended and used for classified and relevant results on the web.

Fig. 2 Repositories of Web Pages

Fig.2. shows the repositories of web pages which contain plain HTML pages and HTML pages associated with ontology.

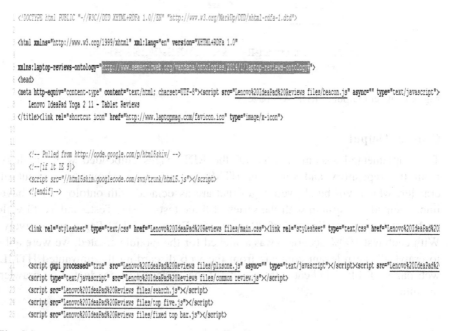

Fig. 3 Screenshot of a Web Page associated to Ontology

Fig.3. Shows an example of a web page   associated with ontology. SemCrawl will crawl these pages associated with ontology and filter out HTML web pages with which no ontology is associated.

5.1   Crawler Evaluation

```
<?xml version="1.0" encoding="UTF-8"?>
<rdf:RDF
       xmlns:dcterms="http://purl.org/dc/terms/"
       xmlns:rdf="http://www.w3.org/1999/02/22-rdf-syntax-ns#">

<rdf:Description rdf:about="file:/E:/projects/RDF/Final/web-
content/Dell%20Inspiron%20Seller.htm">
       <dcterms:title xml:lang="vi">Dell Inspiron
Sellor</dcterms:title>
</rdf:Description>

<rdf:Description
rdf:nodeID="noded3e3c44b27e282f69cdfd72b2b36f0bc">
       <rdf:type rdf:resource="http://schema.org/Product"/>
       <name xmlns="http://schema.org/Product/" xml:lang="vi">
                       Dell Inspiron 15 3521 Laptop (3rd Gen Ci3/
4GB/ 500GB/ Linux)           </name>
</rdf:Description>

<rdf:Description rdf:nodeID="node9285a2090813a4561507dae4945778">
       <rdf:type rdf:resource="http://schema.org/Offer"/>
```

Fig. 4 Crawler Output of RDF Webpage

Crawler Output

The implemented system crawled all the RDF pages embedded with ontology from the repository and filter out HTML pages. Fig.4. shows the output of crawler, which will be all web pages that are associated with ontology. We have implemented our system with the series of three tests Test1, Test2 and Test3 with each test taking a collection of 20 pages, 50 Pages and 100 Pages respectively, With each test 100% accuracy was achieved for the module created, we were able to crawl all the relevant contents from the web discarding the unwanted HTML web pages. Table 2 indicates the results of the various test conducted on crawler module.

Table 2 Series of Test conducted on Repositories for Crawler Module

Test	Repositories	Repositories	Crawler Output
Test 1	20 Pages	RDF Web Pages	20 Pages
Test 2	50 Pages	RDF,OWL,HTML	38 Pages with filtered out 12 HTML pages
Test 3	100 Pages	RDF,OWL,HTML	85 Pages with filtered out 15 HTML pages

5.2 Parser Evaluation

Parser modules have been implemented in java using eclipse framework and jena API library. Jena is convenient toolkit to manipulate RDF models for developing application within semantic web [3]. Parser module extracts triples from the semantic database repositories. Triples are in the form <Subject, Predicate, Object>. Number of triples depends on the vocabulary of our ontology. Output of triples extracted of parser module developed by us has been specified in Table 2. and Fig. 5 shows the output of parser module showing the triples relation<S, P, O>.Table 3 indicates the result of triples discovered by parser module.

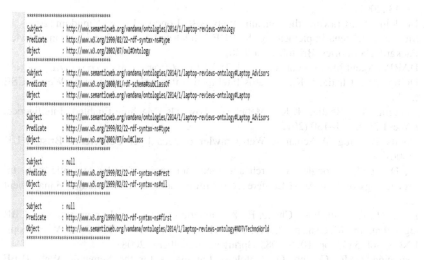

Fig. 5 Console output of parser module extracting Subject, Predicate, Object<S, P, O> Triples

Table 3 Triples discovered by Parser Module

Number of Web pages crawled	20
Number of Triples discovered	146 Triples

Crawler and parser module were executed on particular repository and RDF triples.

6   Conclusion

This paper presented a detail insight into the various differences between crawling in the traditional web as compared to crawling in the semantic web documents. Also the technique with which semantic data can be crawled, for later indexation and classification is proposed. In this research, crawler framework is implemented for harvesting web pages associated with ontology and creating semantic web based knowledge base. In future work further study on the different indexing mechanism of the ontologies associated with the web pages will be done. This paper is concluded with that there is great potential for research in this area regarding how these documents can be indexed which is a very innovative concept.

References

[1] Berners-Lee, T., Hendler, J., Ora, L.: The Semantic Web. Scientific American 284(5), 34–43 (2001)
[2] Biddulph, M.: Crawling the Semantic Web. BBC London, United Kingdom (2003)
[3] McBride, B.: Jena implementing the RDF Model and Syntax Specification. Hewlett Packard laboratories, Bristol, UK (2000)
[4] DARPA Agent Markup Language (2012), http://www.daml.org/language/
[5] Dhingra, V., Bhatia, K.K.: Towards Intelligent Information retrieval on Web. IJCSE (2011)
[6] Dhingra, V., Bhatia, K.K.: Metadata: Towards Machine-Enabled Intelligence. IJWesT 3(3), 121–130 (2012)
[7] Dodds, L.: Slug: A Semantic Web Crawler in Jena User Conference Bristol, UK (2006)
[8] Li, D., et al.: Swoogle: A search and metadata engine for the semantic web. In: Proceedings of 13th ACM Conference on Information and Knowledge Management (2004)
[9] Dong, H., Hussain, F.K., Chang, E.: A semantic crawler based on an extended CBR algorithm. In: Meersman, R., Tari, Z., Herrero, P. (eds.) OTM 2008 Workshops. LNCS, vol. 5333, pp. 1076–1085. Springer, Heidelberg (2008)
[10] Asunción, G.-P., Corcho, O.: Ontology Languages for the Semantic Web. IEEE Intelligent Systems Journal (2002)
[11] Andreas, H., Hannes, G.: On Searching and Displaying RDF Data from the Web. In: 2nd European Semantic Web Conference (ESWC 2005), Heraklion, Greece (2005)
[12] Hendler, J., Berners-Lee, T.: From Semantic Web to Social Machine. Artificial Intelligence 174(2), 156–161 (2010)
[13] Vishal, J.: Ontology Based Information Retrieval in Semantic Web. International Journal of Information technology and Computer Sceince, 62–69 (2013)

[14] Annett, M., Ronny, W., Klaus: Searching Community-built Semantic Web Resources to Support Personal Annotation. In: Proceedings of Bridging the Gap between Semantic Web and Web 2.0, Austria (2007)

[15] Van de Maele, F., Spyns, P., Meersman, R.: An Ontology-Based Crawler for the Semantic Web. In: Meersman, R., Tari, Z., Herrero, P. (eds.) OTM 2008 Workshops. LNCS, vol. 5333, pp. 1056–1065. Springer, Heidelberg (2008)

[16] Staab, S., Apsitis, K., Handschuh, S., Oppermann, H.: Specification of an RDF Crawler (2004)

[17] World Wide Consortium RDF Primer,
     http://www.w3.org/TR/rdf-primer/

OWLSGOV: An Owl-S Based Framework for E-Government Services

Hind Lamharhar, Laila Benhlima, and Dalila Chiadmi

Abstract. The development of e-Government services becomes a big challenge of many countries of the world. However, in a distributed environment such as the e-government area, different interactions are made between heterogeneous systems. Therefore, a system that enables developing, integrating, discovering and executing these services is necessary. For this purpose, we present in this paper an approach for developing efficiently the e-government services based on semantic web services (SWS) technology and Multi-Agent Systems (MAS). In fact, the SWS enrich web services with semantic information (meaning) to facilitate the discovery, integration, composition and execution of services. The MAS enable building an environment in which the public administrations can publish their services, users (e.g. citizens) can express their needs and services can be discovered. In this paper, we present our framework for semantic description of e-government services based on SWS and on OWL-S framework in particular. We present as well the architecture of a MAS, which allows improving the dynamic usage processes of e-government services such as integration and discovery.

Keywords: Semantic web services, e-Government, OWL-S, Multi-Agent Systems (MAS).

1 Introduction

The e-government (electronic government) area aims offering services to citizens. Indeed, the objective of this area is to provide e-services in an integrated, transparent, and efficient manner according to the needs and expectations of business

Hind Lamharhar · Laila Benhlima · Dalila Chiadmi
Mohammed V-Agdal University,
Ecole Mohammadia d'ingénieurs (EMI),
Department of Computer Science, Av Ibn Sina, BP 765, Rabat, Morocco
e-mail: hd.lamharhar@gmail.com

© Springer International Publishing Switzerland 2015   225
R. Buyya and S.M. Thampi (eds.), *Intelligent Distributed Computing*,
Advances in Intelligent Systems and Computing 321, DOI: 10.1007/978-3-319-11227-5_20

and citizens. Currently, many e-services are delivered separately by several public administrations. However, these services are developed without taking into account the multiplicity of concepts, services, business rules and actors involved in administrative proceedings. That has resulted in difficulties for integrating these services and cooperating between involved actors. To overcome these problems, we are interested in improving the description of public services by using semantic technologies; in particular, semantic web services (SWS) [4]. The SWS are the composition of two technologies: Semantic Web [3] and Web Services. In fact, this technology enriches the web services description by additional semantic information, which enables improving the automation level of usage processes of services such as integration, discovery and composition.

Although, the added value of the application of SWS technology in e-government area, our comparative study of numerous researches and projects applying these technologies, has shown that there are still potential problems that must be addressee such as interoperability and integration issues of SWS. Therefore, an e-government solution based on SWS requires:

- good identification and description of SWS through an enhanced framework for semantic description of public services;
- high automation level of different services tasks for cooperating efficiently between providers and users through intelligent mechanisms and tools.

For this purpose, we present in this paper a framework for developing e-government services. In this context, we have developed a system for managing these services. Our system is based on intelligent agents for discovering and integrating the e-government services.

Our framework of SWS is based on a semantic meta-model that represents multiples types of e-government knowledge such as : public services, domain concepts manipulated by these services, usage situations of services, and the relationships between them that we called service usage contexts [5]. The last both knowledge "usage situations" and "usage contexts" represent a cognitive semantic about services. Thus, our meta-model enables developing an e-government knowledge model shared between involved actors, ensures understanding and interpretation of these concepts in a unified manner, and facilitates the research, integration, discovery, and dynamic composition of services required for an e-government process. For this purpose, in our work, we have used OWL-S (Ontology Web Language for Services) framework [13], a set of referenced models, ontologies, and conceptual structures (CSs) [18]. OWL-S ontology enables semantic description of services and CSs enables modeling the service usage situations and contexts of public services. In addition, we have enriched OWL-S by additional elements for describing the governmental features of public services (e.g. legal and organizational). As result, we have built up an e-Government service ontology: *OwlsGov*.

Furthermore, we have developed an intelligent system for discovering and integrating dynamically the *OwlsGov* services. The architecture of this system is composed of a set of layers which allow users to express their requests, to discover and to perform the required services and public authorities to develop and manage their proper services and domain ontologies. To implement this architecture, we have used numerous semantic technologies such as: Pellet [16] for semantic reasoning about OwlsGov services descriptions and domain ontologies and JADE [7] for implementing the intelligent agents. We have indeed developed a set of agents which permit integrating and discovering the public services taking into account their usage situations and contexts. These agents enable thus to build a multi agent system (MAS) in which users and providers can interact easily. Moreover these agents can integrate and discover public services taking into account their usage situations and contexts.

The rest of this article is organized as follows: we present, in section 2, our OwlsGov ontology, next in section 3 and 4; we present respectively our architecture for developing, discovering service, etc. In order to show the feasibility of our architecture, we present in section 5 an application in the context of the e-customs area. Finally, in the last section, we conclude and present our future work.

2   OwlsGov Ontology

Based on our comparative study of SWS frameworks [10], we have chosen to develop our OwlsGov ontology as an extension of OWL-S ontology. We have added various extensions to this ontology in order to represent the specific governmental features and the service usage situations and contexts ontology. In the next, we first give an overview of OWL-S ontology (Section 2.1) before detailing our OwlsGov ontology (Section 2.2). For services development, we have developed an appropriate methodology (Section 2.3).

2.1   OWL-S Ontology

We have adopted OWL-S Ontology to describe the public services semantics for many reasons: First, it is a very popular language, second, it is built on OWL "Ontology Web Language" which is a recommendation of W3C, and third it is characterized by its flexibility to be adapted and finally, it fits best the fact that public services are often business process through its process model. Furthermore, OWL-S proposes an upper ontology that provides three kind of knowledge about service namely: "ServiceProfile", "ServiceModel", and "ServiceGrounding" the first is characterized by the question "What does the service provide", the answer presents the service functionality and other non functional properties. The second element is characterized by the question "How does service work", the answer describes the Service behavioral aspect. The third is characterized by the question "How to access to service"; the answer gives the concrete service that supports Service [13].

2.2 *OwlsGov Ontology*

Our approach aims to develop a semantic model for e-government domain without changing the core ontology of OWL-S as shown in figure 1. For this purpose, we have developed first our specific conceptual model for public services [8]. It consists of *"GovService"*, *"GovServiceProfile"* and *"GovServiceProcess"* that extend respectively the OWL-S elements: Service, Profile and Process. Consequently, we have incorporated the specificities of e-government domain through using some available metadata and ontologies developed within other projects [17]. For example, we have used *"vcard Ontology"* for semantic description of persons and addresses. We also used *"LegalOntology"* and *"OrganizationalOntology"* developed within OntoGov project.

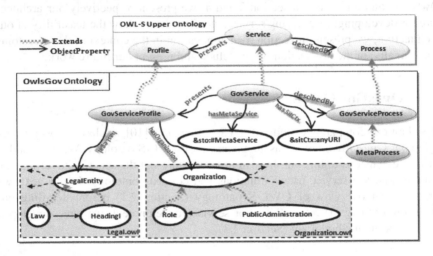

Fig. 1 OwlsGov Conceptual Model

We have enriched the both elements: *GovService* and *GovServiceProfile* by a set of concepts and relations.

Since *"the OWL-S Profile allows the description of a host of properties that are used to describe features of the service"* [13], we decided to integrate the both extensions, **LegalEntity** and **Organization**, which are governmental properties in *GovServiceProfile*. Indeed, the **LegalEntity** concept specifies the legislative characteristics that govern a public service and the **Organization** concept specifies the organizational structure that manages a public service. These both concepts are specified in separate ontologies, *LegalOntology* and *OrganizationalOntology*. The concepts of these ontologies will be instantiated according to the specific regulations and resources, employees and roles of each public administration. To link these concepts to *GovServiceProfile* of a particular service, we have added two relationships **hasLegal** and **hasOrganization**, as shown in figure 1.

On the other hand, *GovService* element enhances the three types of OWL-S knowledge (profile, process, and grounding) by a cognitive knowledge. To integrate this type of knowledge, we have added two concepts, **MetaService** and **SitCtx.**

- **MetaService** is an abstraction of the common knowledge of several services. As example, we consider the calculation services of duties and taxes of e-Customs domain such as VAT and Import Duty (ID) calculation service. Both service share similar functionalities (static) such as their inputs (ex. tax rate) and outputs (ex. amount) but they differ in their dynamic functionalities or applied regulations (e.g. the ID is applied to take into account the country from which commodities are imported).
- **SitCtx** describes the different possible situations (Sit) of a service and their usage contexts (Ctx). To explain this, we consider the example of "Customs Clearance of Goods process". This process can have several situations depending on the type of imported "Commodity" such as "Vehicles" or "Animals". Both processes integrate different services from different public administrations as follow: the customs clearance of vehicle integrates services of transport administration (e.g. vehicle registration) while the customs clearance of animals integrates services of agriculture administration (e.g. sanitary control).

To associate these concepts to their *GovService*, we have added two relationships **hasSitCtx** and **hasMetaService**, as shown in figure 1. We note that, the **SitCtx** knowledge is specified through an external ontology and we have used conceptual structure [18] and conceptual graph [19] to implement this ontology. The technical detail of this part is beyond the scope of this paper.

2.3   OwlsGov Development Methodology

We have developed a methodology for SWS development. [10, 11], which consists of five steps as follow, cf. Figure 2:

Fig. 2 OwlsGov development methodology

- **Step 1:** Requirements Capture is to identify the different types of user (e.g., citizens, businesses, employees and managers of the administrations); needs and required information and services.
- **Step 2:** the Required Ontologies Development is to develop the necessary ontologies such as: domain ontology (CustomsOntology), legal and organizational ontology;

- **Step 3:** Technical Specification Identification consists of developing the OWL-S services with their technical functionalities according to OwlsGov model. We note that, our approach is not limited to OWL-S ontology specification. Indeed, we can use other SWS ontologies.
- **Step 4:** "Governmental Features Identification" is to instantiate the legal and organizational ontologies by each PA with their own governmental characteristics and then to enrich the developed services by these characteristics;
- **Step 5:** "Service Usage Situations and Contexts" is to develop and enrich OwlsGov services with their appropriate **SitCtx** ontologies.

Until now, we have presented our approach for semantic description of e-government services through OwlsGov. Next, we present an intelligent system for discovering and integrating dynamically theses *OwlsGov* services.

3   OwlsGov Management System Architecture

The architecture, shown in Figure 3, represents our OwlsGov Management System (OwlsGovMS). It aims to facilitate the development, the publication, the retrieval and the discovery tasks of e-government services as *OwlsGov* service descriptions. This architecture integrates different types of providers such as public administrations and other organisms. For achieving a specific user task, this architecture supports a set of ontologies that describe the domain concepts and services (OwlsGov) and a set of mechanisms for achieving a specific user request or publishing and integrating services of public administrations.

To build the OwlsGovMS system, we have used the three layers proposed by DIP (Data, Information and Process Integration) project for SWS [6] as follow: User Application layer or front office, Service Provider layer or back-office and the Middleware layer or broker as shown in Figure 2. We detail these layers in the following sub-sections.

Service Provider Layer: This layer composed, in most cases, of public administrations (PA) which develop their own services according to *OwlsGov* ontology model. For this purpose, we propose two approaches for developing PA OwlsGov services, Bottom-up and Top-down. In the first approach, OwlsGov services are developed from the existing web services and in the second approach, services are developed from scratch. In both cases, the services provider must follow the methodology presented in previous section. We note that, the majority of existing web portals of public administration are developed as e-services which are not primarily web services. So, we propose to adapt them to be accessed as web services using auxiliary software agents.

To simplify the development task of *OwlsGov* services, we have implemented a graphical editor and defined a set of tools for developing and updating the associated domain ontologies.

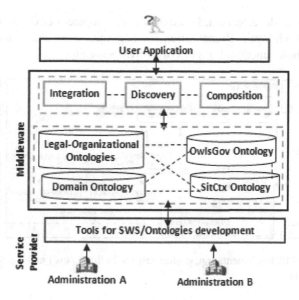

Fig. 3 OwlsGov Management System Architecture

Middleware Layer: The middleware is the core of this architecture and represents the component that connects the back office to the front office. It provides the necessary intelligent functionalities for integrating, cooperating and discovering services. It contains also of a set of repositories for ontologies storage. The interactions between the different middleware components are implemented through a multi agents system (MAS). This latter is composed of agents such as *Client Agent* for query processing, *Discovery Agent* for matching between a given requests and OwlsGov services, and *Integration Agent* for services processing and classification, etc.

User Application Layer: The last layer is represented through a web application through which users can express their queries. This layer consists of several graphical interfaces, which allow collecting the appropriate information for improving service discovery process. To this end, in our system, we have developed a Client Agent for capturing, processing and sending information to the middleware, and then displaying the results (required services and information) in a user graphical interface.

4 OwlsGovMS Implementation

The prototype proposed consists in the implementation of OwlsGovMS architecture. It is based on a set of software components and ontologies, as shown in figure 4. We used the eclipse development environment [20], JADE (Java Agent DEvelopment Framework [7] for implementing agents, PDE (Plug-in Development Environment) for implementing a plug-in application for

OwlsGov services development. Based on this development environment, we have developed the both levels, **Presentation** and **Service**, which enable the implementation of the whole functionalities of OwlsGovMS architecture.

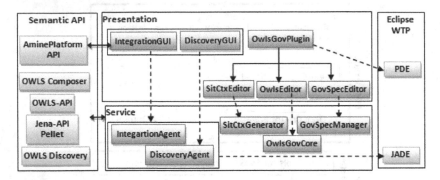

Fig. 4 OwlsGovMS Implementation, dashed arrows indicate (use) and black arrows indicate a composition

4.1   Presentation Level

This level implements the main functionalities that will be used by of the "**User Application Layer**" and "**Service Provider Layer**" of the architecture. It proposes different graphical editors and user interfaces that will be used in by both: public administrations and users. Public administration, represented as service providers, use these interface to develop their own ontologies and *OwlsGov* descriptions. Users, represented by the Application layer of OwlsGovMS architecture, can submit their request through these interfaces. Thus, the Presentation level consists of two main parts: the first part is developed based on plug-in technology, we call it **OwlsGovPlugin.** The second part is developed as a Graphical User Interface for agents' execution.

OwlsGovPlugin is developed as an eclipse Plug-in and uses OWL-S API [14], OWLS Composer [15], etc. It consists of a set of packages as follows:

- "actions" package consists of a set of action such as *ConvertWSDL* that transforms a WSDL descriptions of web services to OWL-S descriptions. To create services from scratch, there is a *CreateService* action. To generate automatically a first version of SitCtx ontology from OwlsGov descriptions, there is a *GenerateSitCtx* action and so on.
- "Generator" package contains all required classes for generating OWL-S ontologies and ontologies SitCtx.
- "editors" package contains editors and various pages that allow to edit the technical specification, governmental features and the usage situations/contexts of OwlsGov services. This package uses the classes defined in both packages generator and actions.

- "wizards" package contains the wizards pages that assist the developers for developing a new *OwlsGov* service.
- "owls-diagram" package of OWLS Composer provides tools to build a static composition plan. We use it to enrich *OwlsGovPlugin* by additional actions such as *GenerateComposition* and runComposition.
- "owls" package provide the main technical functionalities.

IntegrationGUI and **DiscoveryGUI** interfaces are developed based on AminePlatform interface. **DiscoveryGUI** is a GUI service discovery agent and uses the SitCtx ontologies. This interface is a sub class *of IntegrateDefinitionFrame*, which allows introducing a user query, executing the discovery agent, displaying the result to user, and finally selecting the required services. **IntegrationGUI** is the interface that allows integrating a new service using his *SitCtx* ontology in a specific Service Integration Ontology (SIO).

4.2   Service Level

The second layer implements the main functionalities of the architecture "**Middleware Layer**" and provides the whole needed services for interact with the middleware components, with the user application, and with the providers. We have identified two types of services. The first kinds of services are related to the development of the OwlsGov services: **OwlsGovCore** manage the OwlsGov attributes; **GovSpecManager** maintains the governmental characteristics; **SitCtxGenerator** is the interface that generates the SitCtx from its OwlsGov description. The second kinds of services represent the MAS part of our OwlsGovMS, in which every agent has a specific frame that displays information about it such as names, actions, and messages. In this paper, we limited to DiscoveryAgent and IntegrationAgent.

DiscoveryAgent is agent that plays the role of an intermediate between a client agent and service agent. It executes the matchmaking algorithm based on the SitCtx ontologies of services as a first stage of discovery process and then based on their technical specification as the second stage. This agent is developed using the dynamic interference process of AminePlatform. **IntegrationAgent** applies an integration algorithm of a new service using the same process. Thus, each IntegrationAgent, according to the service usage, integrates the new service into an appropriate SIO ontology. Thus, this agent can call the DiscoveryAgent to find the similar services of the new service and then integrate it to the SIO of these similar services.

4.3   Semantic API

The use of semantic ontologies required appropriate semantic API. In this context, numerous tools and semantic java API are developed in literature. To minimize the cost and the time of development we have reused some of them. We present briefly these tools and semantic API with a brief explanation about how we used them in the context of our work.

- **OWL–S API [14]** is a Java API developed by Mindswap to create, read, write, and execute the OWL-S services descriptions. We use this API to implement the OwlsGov conceptual model. According to this model, we have used the three Interfaces namely, *OwlsService, OwlsProfile* and *OwlsProcess* to implement respectively the *GovService, GovServiceProfile* and *GovServiceProcess,* and then we have enriched them with additional extensions of OwlsGov model such as: Organization, Legal, SitCtx and MetaService.
- **OWL-S Composer [15]** is an eclipse plug-in developed by FORMAS (Research Group on Semantic Applications and Formalisms) in Federal University of Bahia. Its main functionalities are the discovery of similar SWS; the composition of SWS under eclipse in a graphic and visual manner. It consists of four plug-ins namely Owls_3.0.0, Diagram, Editor and Edit. We have used mainly the Diagram plug-in to help the developers to construct a static composition of services in case of administrative procedures.
- **OWL-API** is a Java API to create, manipulate and serialize OWL ontologies. We used this API to exploit in particular the domain ontologies such as the Legal and Organizational concepts that manage a given OwlsGov service and the SitCtx associated to it and so on.
- **Pellet [16]:** is an open source OWL Reasoner for OWL ontology-based data management applications. We have used it for multiple purposes such as: connecting to a specific OwlsGov service ontology through a graphical user editor that loads information about this service (e.g. inputs/outputs). We have also used Pellet for implementing the discovery process.
- **AminePlatform [1]:** is an open Java platform to create, edit and modify ontology. This platform provides graphical interfaces for a direct interaction with users and a set of API for java programming. In the context of our work, we have used some of its classes such as: *Ontology*, CG (Conceptual Graph) and CS (Conceptual Structures) to develop the SitCtx ontology of *OwlsGov* service; the *MemoryBasedInferences* class for services integration, and *AmineJadeMAS* to develop a discovery agent that uses the SitCtx for discovering services.
- **OWL-S Discovery [2]** is a tool developed for OWL-S services discovery. It uses a hybrid algorithm of two-step (functional and structural). We use this tool to implement the second step of the discovery process. To this end, we have added the owlsDiscovey.jar API to the *OwlsGovMS* Classpath.

5   E-Customs Case Study

Among the main activities of e-Customs area is allowing the "home use" of imported commodities. The scenario of goods customs clearance (GCC) is complex and composes of a set of information about commodities, taxation and required administrative documents, which must be specified to facilitate administrative procedures of customs clearance to citizens. For this purpose, we have applied our **OwlsGovMS** for developing semantically, data, and services in this area.

The GCC process as shown in figure 5 is shared between a set of involved PAs, such as Customs Administration (e.g. duties and taxes), Agriculture (e.g. clearance of Animals), and Transport (e.g. clearance of Vehicle). CCG is composed of various services such as "CommodityClassification", "RequiredDocument", "ServiceControl" and "ComputeTax". Each service describes one task of the CCG process: "CommodityClassification" for retrieving products codes, "ComputeTax" for calculating D&T, "RequiredDocument" for researching the list of documents as well as their responsible administrations, and "ServiceControl" for commodities control according to involved PA.

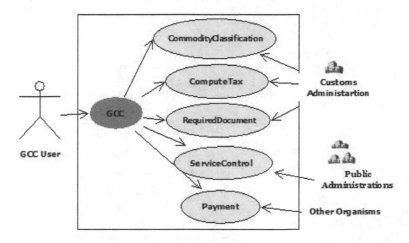

Fig. 5 OwlsGovMS Goods Customs Clearance case study

We have used **OwlsGovMS** to implement the e-Customs services. The detail specification of this case study is given in our work [12]. In this paper, we focus only on *CommodityClassification* service that enables getting a commodity code from a given user request. We illustrate in figure 6 how we use **OwlsGovPlugin** editors to develop this service. The *"GovSpecOntology Tab"* enables to specify the domain ontology that is « *CustomsOntology* ». This latter is developed within our work to represent the e-Customs knowledge base. *CommodityClassification* service is managed by "CustomsAdministration" and governed by CustomsLaw. The *"Owls Upper ontology Tab"* enables to specify the functional features of this

service such as Commodity as the service input and Classification as the output.
The *"Usage Sit/Ctx Ontology Tab"* enables to specify how CommodityClassification is used through modeling its relationships and situations (Sit#).

Fig. 6 OwlsGovMS/OwlsGovPlugin for CommodityClassification service

To illustrate the **DiscoveryAgent**, we consider a user request that is looking for
the code classification of a good of an animal type: "what is the classification of
Animal?" This request is captured through a user interface and then processed automatically by the OwlsGovMS system. Figure 7 illustrates the execution of discovery process through GUI interfaces.

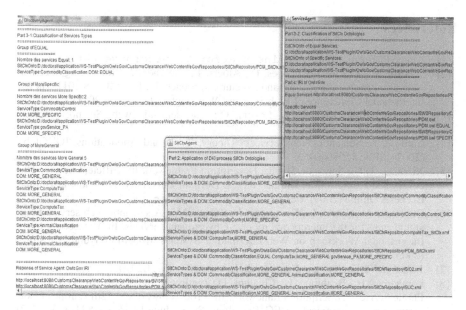

Fig. 7 OwlsGovMS/DiscoveryAgent execution

6   Conclusion

For an e-Government solution based on SWS technology, we have presented in this article a framework for semantic description of public services, the OwlsGov ontology. We have presented as well a system for developing and managing these services that we have called OwlsGovMS. To implement this architecture, we have identified and used several semantic technologies and tools such as ontologies, OWL-S ontology, multi-agent systems, etc. Thus, for the achievement of data and services interoperability, this system offers various components for an efficient use of these ontologies.

As result of this work, we have developed a "graphical user interfaces", which can assist the public administrations for the development of their own services and information. OwlsGovMS enable as well developing an e-government portal which assists users to search and discovering the required services.

As part of future works, we aim to use our model to develop business processes, taking into account the different interpretations and understandings from different perspectives of users. For that, we aim to enhance the existing "LifeEvent" model for citizens or Business-Episode model for enterprise through using our dynamic discovery and integration agents. These models will be used to integrate, discover and compose services according to user's point of view in a specific domain application such as e-Customs domain.

References

[1] Amine Platform, http://amine-platform.sourceforge.net/
[2] Amorim, R., Claro, D.B., Lopes, D., Albers, P., Andrade, A.: Improving Web service discovery by a functional and structural approach. In: IEEE ICWS 2011-The 9th International Conference of Web Services, Washington, D.C. (July 2011)
[3] Berners-Lee, T., Hendler, J., Lassila, O.: The semantic web. Scientific American 284(5), 28–37 (2001)
[4] Cardoso, J.: Semantic Web Services: Theory, Tools and Applications. IGI Global (2007) ISBN: 159904045X
[5] Xiaofeng, D.U.: Semantic Service Description Framework for Efficient Service Discovery and Composition, Durham theses, Durham University. Available at Durham E-Theses Online (2009), http://etheses.dur.ac.uk/111/
[6] Gugliotta, A., Cabral, L., Domingue, J., Roberto, V.: A semantic web service-based architecture for the interoperability of e-Government services. In: Proceeding of the International Workshop on Web Information Systems Modeling, Sydney, Australia (2005)
[7] Java Agent DEvelopment Framework, http://jade.tilab.com/
[8] Lamharhar, H., Benhlima, L., Chiadmi, D.: Incorporating Context in OWLS-Based Public Services Description Framework. In: Proceedings of the 12th European Conference on e-Government, ECEG, Barcelona, Spain, June 14-15, pp. 834–843 (2012)
[9] Lamharhar, H., Chiadmi, D., Benhlima: How semantic technologies transform e-government domain: A comparative study and framework. Transforming Government: People, Process and Policy 8(1), 49–75 (2014)
[10] Lamharhar, H., Chiadmi, D., Benhlima, L.: A comparative study on Semantic Web Services frameworks. In: Proceedings of the Third International Conference on Web and Information Technologies, ICWIT, Marrakech, Morocco, pp. 449–460 (June 2010)
[11] Lamharhar, H., Chiadmi, D., Benhlima, L.: Moroccan e-government strategy and semantic technology. In: Government e-Strategic Planning and Management. Public Administration and Information Technology, vol. 3, pp. 323–343. Springer Science and Business Media, Publisher of the Public Administration and Information Technology (2014), http://www.springer.com/series/10796
[12] Lamharhar, H., Kabbaj, A., Chiadmi, D., Benhlima, L.: An e-government knowledge model: 'e-customs' case study. An International Journal of Electronic Government 11(1/2), 59–82 (2014)
[13] Martin, D., Burstein, M., Hobbs, J., Lassila, O., McDermott, D., McIlraith, S., Narayanan, S., Paolucci, M., Parsia, B., Payne, T.R., Sirin, E., Srinivasan, N., Sycara, K.: OWL-S: Semantic Markup for Web Services (2004)
[14] OWL-S API, http://on.cs.unibas.ch/owls-api/
[15] OWL-S Composer, http://sourceforge.net/projects/owl-scomposer/
[16] Pellet home, http://pellet.owldl.com/
[17] Peristeras, V., Tarabanis, K.A., Goudos, S.K.: Model-driven eGovernment interoperability: A review of the state of the art. Computer Standards & Interfaces 31(4), 613–628 (2009)
[18] Sowa, J.F.: 'Conceptual Structures': Information Processing. Mind and Machine. Addison-Wesley, London (1984)
[19] Sowa, J.F.: Conceptual Graphs for Representing Conceptual Structures (1992), http://www.jfsowa.com/pubs/cg4cs.pdf
[20] Web Tools Platform (WTP), http://www.eclipse.org/webtools/

A Survey on Reduction of Load on the Network

Rajender Nath, Naresh Kumar, and Sneha Tuteja

Abstract. The following research paper emphasizes on ever increasing load on networks vitally due to the presence of web crawlers and inefficient search mechanisms. Consequently, a lot of research has been done already but no feasible solution has been found yet. The following research paper tries to find out the possible loopholes by surveying previous researches so that one can come up with a more practicable and workable approach to lessen the load on the network.

Keywords: WWW, Web Crawler, Network Bandwidth, Software agents, URL, Hypertext, Domain.

1   Introduction

WWW commonly known as World Wide Web [1] is client server architecture having hyperlinked and heterogeneous information that includes audio, video, text and images [2]. HTTP and HTML plays a very important role in presenting the hyperlinked document. According to an estimation of [3], WWW currently contains more than 4.16 billion web documents. Due to abundance of documents that are available on the web, a search tool called search engine is used to search the information on the web. A search engine containing a widespread database of documents can provide more powerful search facilities [4]. For this, a search engine creates its database with the help of web crawlers [5]. According to a study of [6], any web crawler can provide more than 3 billion documents for indexing but at the same instance the study of [7] concludes that apparently it is only 16% of the entire web. So, a search engine with large database has to consider two

Rajender Nath
DCSA,
Kurukshetra University, Kurukshetra

Naresh Kumar · Sneha Tuteja
CSE Department,
Maharaja Surajmal Institute of Technology, New Delhi

© Springer International Publishing Switzerland 2015   239
R. Buyya and S.M. Thampi (eds.), *Intelligent Distributed Computing*,
Advances in Intelligent Systems and Computing 321, DOI: 10.1007/978-3-319-11227-5_21

issues. The first issue is crawling strategy which decides the next page to be downloaded. The second issue is to maximize the download speed with minimum utilization of resources [8].

In this paper, a survey on existing works to resolve the above said issues has been done. This survey contains the articles from 1999 to 2013 and a comparison is provided to showcase the scenario of the existing systems. This paper is assembled as: section 2 which describes the related work. On the other hand section 3 describes the general architecture of a web crawler followed by section 4 that states the challenges faced by the web crawler and section 5 concludes the paper.

2   Related Work

A parallel domain focused crawler approach has been proposed in [9]. In this, FCE or frequency change estimator was used in order to calculate the frequency of change of the web documents assigned. It calculates the frequency in the following manner-

F=-1 (for URLs that were being traversed for the first time)
F= 0 (for web pages having rapid frequency change, so the frequency parameter could not be applied here)
F= N (N>0, where N is the total number of days since last page change occurred)

Further the Com (comparator module) compares the web pages on the basis of ASCII Count of the new pages with the old ones that had been downloaded. If the ASCII Count matches, the new page was discarded; else the old page was updated. The proposed approach lowers the load on the network to a large extent. The load was further reduced on compression. Moreover network bandwidth was preserved by reducing the traffic of the network. The proposed system filters off the unmodified pages locally at the remote sites. In Mobile crawlers approach, based on filtering off non modified pages that was proposed by [10], comparator module compares the statistics collected from the old database file module with the statistics of the corresponding web page collected from the analyzer module at the remote site. If the last altered date did not match, the page was sent to search engine for listing. The paper also proposed a way to check whether the retrieved page had been modified or not in case last modified date was not present. All those web pages whose last modified date was not present were sent to AM (analyzer module). AM scans the pages one at a time and counts the number of URLs. If the count varies, the pages were indexed and if the URL count was same, another comparison regarding the number of keywords present in the web page was done. In the case of conflict occurs the pages were sent to search engine for indexing, else they were discarded. In [11] a compressing technique had been stated in order to reduce HTML network traffic. Original HTML files were shrunk, encoded, and then compressed to form reduced sized HTML files called zHTML files that were stored on the server site. Also a plug-in program had been proposed so that the users do not find any difference when viewing HTML documents with reduced size as compared to the original ones.

Further, more Literature survey on reduction of network load is presented in tabular form:-

Table 1 Papers exhibiting different techniques for curtailment of load

Ref no.	Year	Author's name	Implementing technique	System used
[12]	1999	Rawn Shah	Web Caching	A web cache stores web resources in anticipation of future requests. It works on the principle that popular resources are most likely to be requested in future.
[11]	2000	Joachim Hammer and Jan Fiedler	Remote page filtering	The proposed system allows the crawler to control the granularity of data it seeks. After retrieving pages, it filters out the irrelevant data retaining only data that has been validated by the search engine.
[13]	2001	Wang Lam and Hector Garcia-Molina	Central Crawler	A single central crawler builds a database of WebPages and provides services for multiple clients. In such a way, clients can alternatively subscribe to the central crawler for the pages they need and only one crawler visits the webpage and distributes the data to their respective clients.

Table 1. (*continued*)

[14]	2002	Ben Choi and Nakul Bharade	Hypertext Compression	Process includes shrinking HTML documents, encoding and compressing these documents in order to reduce bandwidth used. Also a browser plug-in program that provides transparency of data for the users was developed.
[15]	2003	Odysseas Papapetrou, George Samaras and Stavros Papastavrou	UCYMICRA	Distributed indexing of the web using migrating crawlers was proposed. Server registers itself to a particular search engine and thereafter migrating crawler was allowed to dispatch itself to the web server.
[16]	2004	George Samaras and Odysseas Papapetrou	IPMicra	Location aware web crawling was suggested. In this, pages were distributed to crawlers so that each page could be crawled from the nearest crawler. This results in downloading the pages faster and with fewer resources.
[17]	2007	Debajyoti Mukhopadhyay, Arup Biswas and Sukanta Sinha	Domain specific Ontology based Crawlers	Weights were calculated so that the more relevant and specific terms will have more weights to it and the terms common to more than one domain should be given less weightage followed by the determination of the validity of the page's content.

Table 1. (*continued*)

[18]	2008	Lefteris Kozanidis	Based on ontology	• Relevance of the page to the ontology topic that has been assigned was calculated • Text nugget that was most semantically close to page's topical content was extracted Topic Relevance was calculated as :- Thematic keywords in page matching t/ thematic keywords in page (where t is ontology topic).
[19]	2010	Achintya Das and Sudarshan Nandy	Hybrid Focused Crawler	Topic related web research was recommended for domain specific search. Crawler crawls only those pages that were considered relevant to the topics assigned to it.
[10]	2011	Rajender Nath and Satinder Bal	Mobile Crawlers	Mobile crawlers compare the web pages on the basis of the number of keywords and the URLs present in the pages with the previously stored data in the old database file system. Based on comparison, it filters off those pages that were not revised since the last crawl.
[9]	2012	Naresh Kumar and Rajender Nath	Parallel Domain Focused Crawler	Uses frequency change estimator that calculates the frequency of change and filters out those pages that had low frequency. Further comparator module compares the pages on the basis of ASCII count with the data of the corresponding page present in the old database file system.

Table 1. (*continued*)

| [20] | 2012 | Niraj Singhal, Ashutosh Dixit, R.P Aggarwal and A.K Sharma | Migrating Crawlers | Crawler Manager at a search engine site deputes migrating crawlers with a list of URLs. Crawlers interact locally, crawls the pages, selects the best page and comes back. Parallel and multiple migrants were used for scalability. |
| [21] | 2013 | R.G. Tambe and M.B. Vaidya | JINI Technology | Mobile crawlers using JINI Technology were sent to the remote servers with the user's queries. Crawlers select the pages that were considered relevant with respect to the application domain, strain the irrelevant data and finally compress the page and move to the next server. After collecting several relevant pages a web crawler returns to the search engine. |

3   General Architecture of a Web Crawler

Web crawler recursively uses GET and POST command to traverse the web pages and further downloading them for search engine repository. To maintain the freshness of repository a crawler needs to revisit the web site repeatedly. Web crawler starts its process of searching by taking the initial URL, downloading the corresponding web page, extracting hyperlinks (if available) from downloaded

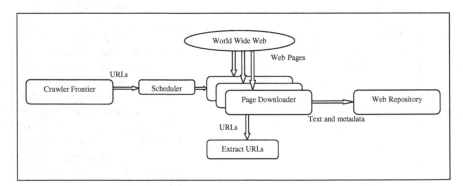

Fig. 1 General architecture of a web crawler [22]

web pages and iteratively repeating this process till it finds all the URLs are downloaded. For further processing the downloaded web pages are indexed at the search engine end.Fig.1 and Fig.2 show the architecture and working of a web crawler.

In summarized way general architecture of web crawlers works as follows [22]:

1. It starts with the initial URL known as seed URL which is further added to the frontier.
2. URLs from the frontier are then visited recursively.
3. Web pages corresponding to each URL are fetched (if robots.txt file allows).
4. Fetched pages are parsed to extract hyperlinks.
5. All the unvisited pages are added to the frontier.

Steps 1 to 5 are repeated till frontier is clear.

3.1   Mechanisms Used in Web Crawling

Different strategies or mechanisms being used for web crawling are as follows [5] [22] [24][26]:-

1. Mobile crawlers: - These are the mobility based crawlers. They crawl the pages using mobile agents. Search engines transfer these crawlers to remote sites where they reside and filter out the extraneous data. They perform continuous examining of all the web documents assigned to them. They help to reduce the load by compressing the documents locally and transferring them across the network [11].

2. Domain specific crawlers: - These crawlers traverse the web pages according to a particular domain. Basically there are two types of domain specific crawlers-

A) Topic Specific Crawler: - A topic specific web crawler crawls the web pages according to the relevant topic. It does not collect all the web pages.

The relevance can be computed in the selected domain. If it is found important the page gets added, else it is discarded.

B) Ontology based crawlers: - Ontology is a shared wordlist or lexicon created for the researchers to denote the type and interrelations between them. Crawlers crawl the pages relevant to a given ontology.

3. Based on coverage area: - Crawlers based on coverage area are of two types:-

A) Unfocused Crawlers: - These crawlers crawl the entire web in order to construct their index. As a result they manage more labor intensive work of creating and refreshing a database with many dimensions.

B) Focused Crawler: - These crawlers limit their function upon a semantic web zone by retrieving pages related to a predefined topic and leaving out irrelevant content or pages [25].

4. Based on Load distribution: - In order to decrease the bandwidth usage and increase the area covered, load is distributed. Depending upon load distribution, crawlers are of two types:-

A) Intra-site parallel Crawler: - Crawlers run on the same local network and exchange information through a high speed interconnection.

B) Distributed Crawler: - Crawlers run in geographically distant locations that are connected through internet. In this several crawlers crawl information simultaneously on different networks.

4   Challenges Related to the Web Crawler

The study of above literature shows various challengeable tasks that persist in the system. Some of them are as follows: -

- In papers [9][10], it has been stated that by comparing URLs and ASCII count, filtering off non modified pages is possible. But the pages are needed to be downloaded in order to calculate size and other measures which results in wastage of bandwidth and other resources.

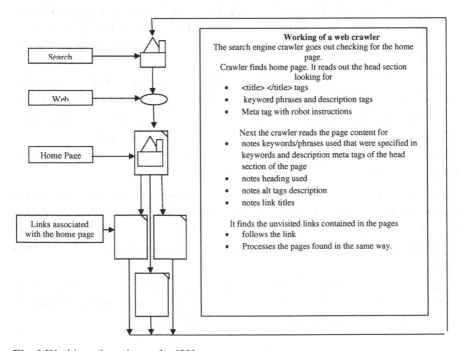

Fig. 2 Working of a web crawler [23]

- The crawlers can degrade the information about the pages in order to reduce the amount of data to be transferred [11].
- In caching based system, [12] the access is entirely decided by the user's request pattern, thus making it arbitrary and covers only a particular set of clients.
- Several issues like interference and attacks on the routes can hinder the proper working of multicast routes used during the distribution of information among clients [13].
- During compression [14] quality of audio and videos is reduced. Loss of fidelity is possible. Moreover an extra plug-in is required for transparency of data
- In [15], it is possible that certain web administrators do not find it useful enough to register themselves to a particular web server.
- In case of ontology based crawlers [17] [18], there could be mismatches between the ontology that hinder the combined use of independently developed ontology.
- In case of focussed crawlers [19], a web crawler can possibly miss a relevant page if there does not exist a chain of hyperlinks that connects one of the relevant pages to the home page.
- In [20] [21], it may be possible that the remotes sites may not allow mobile crawlers to reside in its memory due to security reasons. Some issues are still needed to be dealt with for achieving more practical and viable ways to reduce the load exerted on the network.

5   Conclusion

As the number of web pages on the web cannot be restricted, this results in the need of new methodologies to improve the search results and increase the coverage of the web which still remain a challenge for many. Therefore, a new web crawler that can overcome all the above listed issues of web crawling should be introduced. In this article authors provide a general architecture of web crawler with its working and several mechanisms that are used for web crawling. The major objective of this review paper is to investigate previous existing works that has been done for reduction of load on the network. As a result, various challenges were found and presented during the literature review which should help a novice to easily understand the challenges present that are needed to be solved in the area of web crawling.

References

[1] Duhan, N., et al.: Page Ranking Algorithms: A Survey. In: IEEE International Conference on Advance Computing Conference, pp. 1530–1537 (2009), E-ISBN: 978-1-4244-2928
[2] http://en.wikipedia.org/wiki/Web_crawler

[3] http://worldwidewebsize.com/
[4] Bhagat, G.S., et al.: Improved Search Engine Using Cluster Ontology. International Journal of Engineering and Advanced Technology IJAET 2(3), 420–425 (2013) ISSN: 2231-1963
[5] Nath, R., et al.: Web Crawlers: Taxonomy, Issues & Challenges. International Journal of Advanced Research in Computer Science and Software Engineering 3(4), 944–948 (2013) ISSN: 2277 128X
[6] Google Inc. Google (September 2003), http://www.google.com/
[7] Lawrence, S., Lee Giles, C.: Accessibility of information on the web. Nature 400(6740), 107–109 (1999)
[8] Prasanna Kumar, J., Govindarajulu, P.: Duplicate and Near Duplicate Documents Detection: A Review. European Journal of Scientific Research 32(4), 514–552 (2009) ISSN 1450-216X
[9] Nath, R., Kumar, N.: A Novel Parallel Domain Focused Crawler for Reduction in Load on the Network. International Journal of Computational Engineering Research 2(7), 77–84 (2012) ISSN 2250-3005
[10] Nath, R., Bal, S.: A Novel Mobile Crawler System Based on Filtering off Non-Modified Pages for Reducing Load on the Network. The International Arab Journal of Information Technology 8(3) (July 2011)
[11] Fiedler, J., Hammer, J.: Using the Web Efficiently: Mobile Crawlers. In: Proceedings of the Seventeenth AoM/IAoM International Conference on Computer Science, San Diego, CA (1999)
[12] Shah, R.: Reduce Network Traffic With Web Caching. IBM Development Work (1999)
[13] Lam, W., Garcia-Molina, H.: Multicasting a Changing Repository. In: Choi, B., Bharade, N. (eds.) 19th International Conference on Data Engineering (ICDE 2003), March 5-8, pp. 877–882 (2002); Network Traffic Reduction by Hypertext Compression. In: International Conference on Internet Computing, pp. 877–882 (2002)
[14] Papapetrou, O., Papastavrou, S., Samaras, G.: UCYMICRA: Distributed Indexing of the Web Using Migrating Crawlers. In: Kalinichenko, L.A., Manthey, R., Thalheim, B., Wloka, U. (eds.) ADBIS 2003. LNCS, vol. 2798, pp. 133–147. Springer, Heidelberg (2003)
[15] Papapetrou, O., Samaras, G.: IPMicra: An IP-address based Location Aware Distributed Web Crawler. In: International Conference on Internet Computing, pp. 694–699. CSREA Press (2004)
[16] Mukhopadhyay, D., Biswas, A., Sinha, S.: A New Approach to Design Domain Specific Ontology Based Web Crawler. In: 10th International Conference on Information Technology (2007)
[17] Kozanidis, L.: An Ontology-Based Focused Crawler. In: Proceedings of the 13th International Conference on Natural Language and Information Systems: Applications of Natural Language to Information Systems, pp. 376–379 (2008)
[18] Das, A., Nandy, S.: Hybrid Focussed Crawler-A Fast Retrieval Of Topic Related Web Resource For Domain Specific Searching. International Journal of Information Technology and Knowledge Management 2(2), 355–360 (2010)
[19] Singhal, N., Dixit, A., Aggarwal, R.P., Sharma, A.K.: Using Migrating Agents in Designing Web Search Engines and Property Analysis of Available Platforms. International Journal of Advancements in Technology (2012)

[20] Tambe, R.G., Vaidya, M.B.: Implementing Mobile Crawler Using JINI Technology to Reduce Network Traffic and Bandwidth Consumption. International Journal of Engineering and Advanced Technology (IJEAT), Volume 2(3) (February 2013) ISSN: 2249 – 8958

[21] Udapure, T.V., Kale, R.D., Dharmik, R.C.: Study of Web Crawler and its Different Types. in 2278-0661 16(1), Ver. VI, pp. 01–05 (2014) ISSN: 2278-8727

[22] http://reload4btech.blogspot.in/2012/02/working-of-search-engines.html

[23] Gupta, S.B.: The Issues and Challenges with the Web Crawlers. International Journal of Information Technology & Systems 1(1)

[24] Mali, S., Meshram, B.B.: Focused Web Crawler with Page Change Detection Policy. In: 2nd International Conference and Workshop on Emerging Trends in Technology, ICWET 2011 (2011)

[25] Olston, C., Najork, M.: Web Crawling. Foundations and Trends in Information Retrieval 4(3), 175–246 (2010), doi: 10.1561/1500000017

[26] Kausar, A., et al.: Web Crawler: A Review. International Journal of Computer Applications (0975 – 8887) 63(2) (February 2013)

Smart Human Security Framework Using Internet of Things, Cloud and Fog Computing

Vivek Kumar Sehgal, Anubhav Patrick, Ashutosh Soni, and Lucky Rajput

Abstract. Human security is becoming a grave concern with each passing day. Daily we hear news regarding gruesome and heinous crimes against elders, women and children. Accidents and industrial mishaps have become commonplace. Computers and gadgets have progressed a lot during past decades but little has been done to tackle the challenging yet immensely important field of physical security of people. With the advent of pervasive computing, Internet of Things (IoT), the omnipresent cloud computing and its extension fog computing, it has now become possible to provide a security cover to people and thwart any transgression against them. In this paper, we will be providing a security framework incorporating pervasive and wearable computing, IoT, cloud and fog computing to safeguard individuals and preclude any mishap.

1   Introduction

The present era is ushered as era of wearable computing. The power of computing has migrated from our homes and offices to our hands, eyes and body. A computational device which is at least as powerful as sophisticated desktops and servers of yesteryears can be worn on head- Google Glass [1], or wrist- Samsung Galaxy Gear [2] or neck – Narrative Clip [3]. Smartphones and tablets have become ubiquitous and inseparable part of our lives. Fitness trackers like Fitbit [4] and Nike + [5] are helping professional as well as amateur fitness aficionados to track their daily workout and monitor their health. Smart clothing [6] incorporates small electronic components within our daily wear for providing various applications like health monitoring, wearable screens etc. Internet of Things (IoT) [7] is a

Vivek Kumar Sehgal · Anubhav Patrick · Ashutosh Soni · Lucky Rajput
Department of CSE and ICT
Jaypee University of Information Technology,
Waknaghat, Solan, H.P., India
e-mail: vivekseh@ieee.org

© Springer International Publishing Switzerland 2015   251
R. Buyya and S.M. Thampi (eds.), *Intelligent Distributed Computing*,
Advances in Intelligent Systems and Computing 321, DOI: 10.1007/978-3-319-11227-5_22

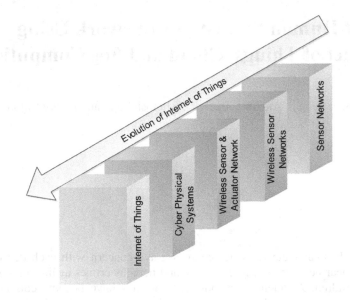

Fig. 1 Evolution of IoT

pioneering concept which aims at connecting all the physical things around us with each other using either traditional internet or geographically restrained wireless ad-hoc networks.

The term "things" here include anything and everything from refrigerators, microwave and cars to smartphones, laptops, servers and supercomputers. IoT devices identify and authenticate each other using communication technologies like RFID and 6LoWPAN. IoT has immense potential of totally changing the way we interact with the world. Potential applications of IoT include smart cities, health monitoring, home automation, smart grids, structural health monitoring etc. According to a recent study by CISCO [8] there will be around 50 billion connected "things" or IoT devices by 2020. Because of the immense potential, IoT is sometimes also proclaimed as the next wave of Internet.

IoT has undoubtedly evolved from Cyber Physical Systems (CPS) which in turn have evolved from Wireless Sensor and Actuator Networks (WSAN) as shown in figure 1. WSAN [9] use wireless sensors (either fixed or mobile) to sense environmental parameters. The data is aggregated from these sensor nodes and intelligent decisions are taken based on some controlling algorithm or predefined logic. The decisions are enforced in real world using in-network actuators or actors. CPS provides a seamless integration between the computation, communication and physical elements of the system [10]. Any cyber system that is interacting with the physical world can be termed as a CPS. Examples of CPS include smartphones, robots, integrated medical devices, UAVs etc. Thus the scope and total entities have increased tremendously after we have moved from WSAN to CPS and ultimately to IOT.

Cloud Computing has metamorphosed from a buzzword a few years ago to cornerstone of contemporary computing in home, offices and enterprises. Cloud computing simply means providing computing as a service rather than as a product [11]. The services provided by cloud are broadly classified as software, platform and infrastructure. The lease and lend model of cloud has provided several benefits to the end users like pay as per usage, anytime-anywhere access, no hardware or software upgradation required, almost limitless storage and computing potential, scalability, low capital cost etc. However, due to network dependent nature of cloud it is not suitable for real time and latency sensitive applications like traffic monitoring, critical health tracking applications in hospitals etc. Fog Computing [12] is a concept which extends the cloud and try to overcome its limitations. Fog computing brings some of the functionalities of cloud to the edge of a network which in turn is beneficial in latency sensitive applications, location-aware applications and geo-distributed applications. In fog computing, we provide computing, storage and communication capabilities to routers, gateways, bridges and other networking components. The result is that some of the simple yet frequent tasks of the cloud can be delegated to fog.

Physical security of people and particularly women is a rising concern [13]. With the proliferation of wearable, mobile and fixed IoT devices, cloud and fog computing, it is now possible to continuously monitor an individual and take proactive and in-time decisions to prevent any mishap. In this paper, we have designed a security framework for people using the above mentioned concepts. Rather than making all the decisions either locally or globally, we have utilized a layered approach having multiple layers with increasing level of sophistication. This ensures that the most relevant decision is taken in least amount of time for each scenario. Although we have designed our framework for physical security of people, however, with slight modifications, it can be applied in other domains as well like smart health monitoring, smart cities etc.

The rest of the paper is organized in six sections. Section one briefly introduces various concepts we will be employing in our proposed work. Section two gives an overview of the work done in related fields. Section three provides a detailed description of the proposed physical security framework for people. It also gives an overview of the unique challenges to be overcome in implementing the proposed framework. Section four gives a case study demonstrating effectiveness of the framework. Section five concludes the paper and gives the future direction of our work.

2   Related Work

IoT and Fog computing are nascent research areas and limited research work has been done to exploit their applications. Although extensive research has been done in cloud and its applications however, few studies specifically focus on interaction with physical world. The amalgamation of aforementioned technologies is nearly unexplored. To the best of our knowledge no existing research work focuses on physical security of individuals. In [14] Jayavardhana Gubbi and his fellow authors have proposed Cloud centric Internet of Things. It is internet centric architecture where objects collect the data. This conceptual framework incorporates ubiquitous devices and their applications. The proposed framework provide support for (1) collecting the data from databases or directly from the sensors, (2) uncomplicated expression for data and logic as operators and (3) if any desirable output is observed, it is passed to output stream which is interacting with the visualization program. They have also proposed participatory sensing that deals with deployment of people centric sensing platforms. People centric sensing reduces the cost of the sensing of environment within the proximity of the user. In [15], Chennying Zeng and his fellow authors have derived the idea of monitoring system for Wild animals using 3G and internet of things. WMNS (Wireless multimedia sensor network) nodes are deployed in forest. When an event of animal entry in the area of the respective node occurs the node triggers automatically and takes the picture which is subsequently updated to the monitoring points. Their approach has utilized Zigbee as the communication technology because of the low power consumption and automatic networking. In [16] Xiaogang Yang and his fellow authors has proposed sensor portal for sensing world in real time named LiveWeb. Their approach has taken in consideration – nature of data, data format, data representation, data management. They have emphasized on requirement of the user like access from anywhere, access at any time and user preference. In [17] Patrick G. McLean has proposed the secure pervasive environment. His approach makes sure that only authenticated person will be allowed to enter in the room. Layered agent architecture has been employed. Agent layer contain (1) security layer, (2) entry, exit and room-context agents, (3) wrapper for low level agents and (4) sensors. Entry agents monitor any person who demands to go inside. They notify security agents which check room list for verification of the person. Entry agents then verify the authentication before permitting person to enter inside. Several researchers have focused on health monitoring of patients. By creating a pervasive computing environment, smart nursing homes can help those elderly people who cannot live unassisted. They can provide assistance to them and monitor all their activities and health [18]. The patients who are having senile mental disorder need to take care of exercise and amount of sunlight they receive. Portable wearable devices can aid patients in such scenarios [19]. They have accelerometer, GPS, gyroscope, illumination sensor etc. which will help the patients in maintaining and monitoring their routine and health.

3   Proposed Approach

3.1   Architectural Overview

Our proposed physical security framework consists of three distinct layers namely, the IoT layer, the fog computing layer and the cloud computing layer as shown in figure 2. The layers are arranged in increasing order of their computational and decisional capabilities. Inter layer entities are connected using myriad communication technologies including wireless communication (Bluetooth, NFC, Zigbee,

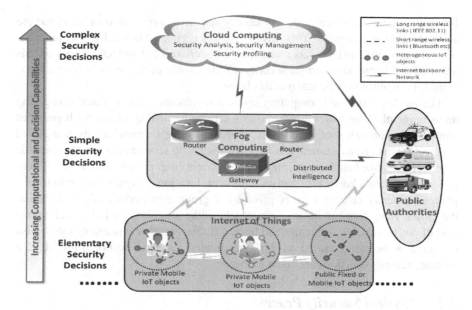

Fig. 2 Proposed Physical Security Architecture

IEEE 802.11 a/b/c/g/n, satellite links etc.), wired communication (optical fibre, Ethernet etc.) or a combination of both. Layering of functionality provides numerous benefits like simplicity, scalability, flexibility, improvability, modifiability etc. Any enhancement to the current layer can be made later on if the need arises. All three layers are connected with public authorities using high speed direct links. In case of any physical threat is detected to a person, the public authorities are immediately alerted along with relevant information.

The lowest layer is the IoT layer. It consists of two main types of entities- mobile IoT objects and fixed IoT objects. An individual will carry several mobile IoT objects with him/her like wearable devices such as smart glasses, fitness tracker, wearable camera, smart clothes etc. along with more traditional mobile computing devices like smartphone and smartwatch. Each IoT object belonging to an individual will be equipped with his unique id along with the unique id of the IoT object. All the IoT objects belonging to an individual will form a private group. All these devices will communicate and cooperate with each other using a wireless ad hoc network. There will be numerous public IoT objects both fixed and mobile installed by government and non-government organizations (NGOs) all around the city. They will be of varying level of sophistication in sensing, computation and actuation. They will be connected with each other wirelessly and will form city wide nexus.

The middle layer is fog computing layer consisting of networking equipment like routers, gateways, bridges etc. augmented with computational capabilities. This layer tends to extend cloud computing to the edge of the network. It comprises of limited computing prowess and autonomy. Apart from doing normal networking functions, many of the real time and latency sensitive applications of the cloud are also relegated to this layer. This layer possesses regional knowledge through various private mobile IoT objects and public mobile/fixed IoT objects. Based on this accumulated knowledge and the security policies, some of the security related decisions for a woman may be taken at this layer.

The top layer is cloud computing layer. It represents a consolidated computing and storage platform which is offered as a service for IoT applications. It provides continuous monitoring of individuals as well as their surrounding from a global perspective. Its primary function is to provide security services but can also provide other services like health monitoring, fitness monitoring, person tracking etc. Security related data like health statistics, current location, neighboring individuals, planned itinerary etc. from every private IoT group gets periodically collected at the cloud. Data from public IoT infrastructure like snapshots of individuals, position of patrol vehicles, vehicular and human traffic, weather conditions etc. is also collected at the cloud. Based on this consolidated data and historical profile of a woman, security decisions are taken for her.

3.2   Physical Security Process

The physical security process starts with smart IoT objects sensing the physical environment. The private IoT objects usually monitor parameters like the location of a person, his health statistics, people with whom he is interacting etc. Public IoT objects monitor more generic parameters like vehicular and pedestrian traffic, structural health, and weather conditions etc. The major difference here is data from private IoT objects is not accessible from outside while data from public IoT objects is publicly available. All IoT objects contain embedded security rules. There are two alternatives for data communication and decision making: 1) Peer to peer approach [20] and 2) Centralized approach

In peer to peer approach, any IoT object that needs to take a security related decision communicates with peer IoT objects in its communication range and gathers the parameters required by the security rule. Based on aggregated parameters, a decision is taken and broadcasted to all nearby IoT objects for subsequent implementation. The above procedure is shown in figure 3(a). There are five private IoT objects and two public IoT objects. If object 1 wants to evaluate security based rule R1 (If B Then A), it queries its peer objects 2 and 4 which in turn forwards the query further to other IoT objects in the group (i.e., device 3) as well as other nearby public IoT objects (i.e., object 6).

The alternative to peer to peer approach is the centralized approach. In this approach, after the initial handshaking and authentication of private IoT devices and nearby public IoT devices, their embedded security rules are mutually exchanged. A group leader is then chosen either arbitrarily or using any leader election

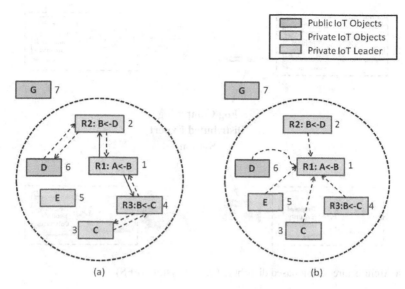

Fig. 3 (a) Peer to peer decision making (b) Centralized decision making

algorithm like Bully algorithm, Ring algorithm etc. Once the group leader is chosen, it then periodically gathers sensed parameters from various IoT objects and takes physical security decisions for the owner of the group. Figure 3(b) demonstrates a system consisting of 5 private IoT objects numbered from 1 to 5 and two public IoT objects 6 and 7 respectively. IoT object 1 is chosen to be the leader of the group. All the security related decisions are taken by object 1 based on relevant parameters gathered from peer devices periodically.

Periodically, data from multiple private IoT groups and public IoT objects gets collected at the networking equipment forming the fog layer for further transmission. Computing capabilities are provided to the equipment like routers, bridges, gateways etc. These devices are sometimes also termed as fog computing nodes [21]. They act as limited capability Distributed Expert System [22][23]. They maintain regional knowledge in the form of a distributed regional security database. They provide low latency, location aware and geo distributed services to various applications including pertinent physical security application. Based on the information received from the entire region, decisions regarding physical security are taken for each private IoT group owner. Fog provides IoT layer greater computing and decisional capability closer to the edge of the network which results in timely and proficient decisions taking into consideration regional perspective and real time requirement of physical security application. The fog layer acts as a limited capability distributed expert system. The architecture of fog based expert system (FES) is shown in figure 4.

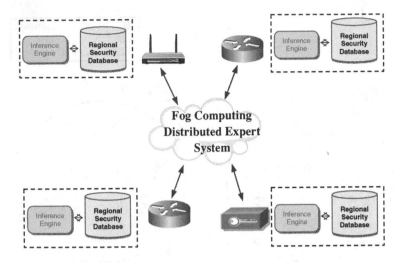

Fig. 4 Architecture of fog based distributed expert system (FES)

Each fog node consists of two essential components- a regional security database (RSB) and an inference engine (IE). The RSB contains various physical parameters sensed by private IoT groups and public IoT devices. The attributes which are not relevant to physical security purpose may be pruned so as to reduce the storage requirements and improve FES performance. The other essential component of FES is IE. IE may employ rule based techniques, neural networks, clustering, support vector machines etc. to infer and deduce conclusions regarding physical security of an individual. IE consults RSB for every decision it makes. In case IE concludes that there is a physical threat to an individual, it immediately intimates the civil authorities along with the concerned individual.

Ultimately sensed data is sent to cloud computing platform which act as a full-fledged cloud based expert system (CES) for detecting any physical threats to a person. The IaaS model of cloud is used for deploying CES. The cloud platform is characterized with almost unlimited and promptly available computing, storage and communicational resources. CES maintain a global knowledge base consisting of security related data of individuals gathered from one or more fog platforms which in turn has been collected from millions of IoT devices both private and public. CES also maintains a security profile of each individual consisting of data like his usual itinerary, health statistics, medical records, etc. The architecture of CES is shown in figure 5. CES works as a sophisticated artificial intelligence based expert system. Apart from traditional inference engine, it also contains two additional sub-components, namely, security analyzer and security manager. The task of security analyzer is to continuously analyze physical security of an individual and consult inference engine in case of suspicion. Security manager manages the overall physical security process and take the necessary preventive and remedial actions. Overall CES acts as an oracle and utilizes its computational prowess and global knowledge to deduce best possible security decisions for individuals. The decisions taken by CES are globally optimal. The complete physical security process of our proposed framework is shown in figure 6.

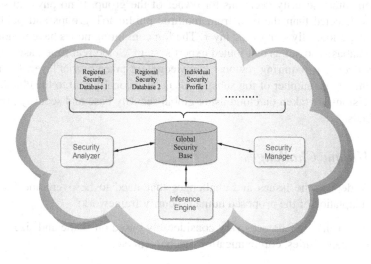

Fig. 5 Architecture of cloud based expert system (CES)

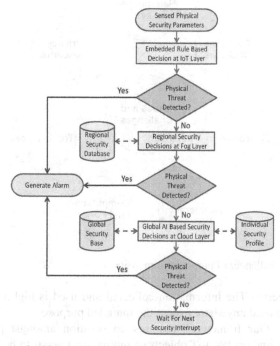

Fig. 6 Flow Diagram for Human Physical Security Framework

   Initially various security related physical parameters are sensed by IoT devices
(both private and public). User's private IoT devices are provided with basic physical security rules. These devices cooperate and communicate with each other to
take elementary security decisions for owner of the group. If no physical security
threat is detected than the data from multiple public IoT groups and public IoT
objects is periodically sent to fog layer. The fog computing nodes have regional security database and act as distributed expert system. Some real time basic physical
security decisions requiring regional knowledge are taken at this layer. Ultimately
data from a large number of fog subsystems is occasionally sent to the cloud. Complex decisions are taken on cloud using global security base and security profile of
individuals.

3.3   Unique Challenges

Figure 7 depicts the issues and challenges that need to be overcome for widespread adaptation of the proposed human security framework:

- Heterogeneity- IoT devices vary considerably based on shape and size, computational capabilities, communication mechanism etc.

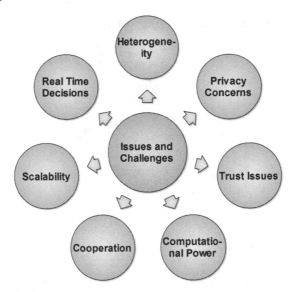

Fig. 7 Issues and challenges of proposed framework

- Privacy concerns- The information collected and used is highly private and it
  should not be used anywhere except the intended purpose.
- Trust Issues- Our framework requires cooperation amongst previously unknown public and private IoT objects so mutual trust needs to be established.

- Computational power- To take accurate decisions with minimum chances of error considerable computational and storage capabilities are required which is generally not available at lower IoT and fog layers.
- Cooperation- In order for framework to function optimally, tight cooperation and coordination amongst multiple layers and numerous entities is required.
- Scalability- Our proposed work requires interconnection between millions of entities. As more number of people will get associated with the framework, number of devices may quickly become unmanageable.
- Real time decisions- Many of the decisions regarding physical security of individuals are real time and even a small delay is unacceptable. Although with the use of fog computing minimizes chances of delay yet it is not fully autonomous and often requires the assistance of network dependent cloud computing.

4   Case Study

In order to assess the effectiveness of our proposed scheme, we consider three scenarios. Suppose there is a person who is been attacked by some attackers. His private IoT group has been augmented with basic security rules, one of which is if the vital body parameters of the IoT owner becomes abnormal then generate an alarm and inform the public authorities. The rule is triggered due to excessive precipitation, faster heartbeat etc. of the victim, and immediately a distress message is sent to the nearby public authorities, along with all the relevant details of the victim including his current location, medical history, social security number etc. In the second instance, consider a scenario in which some robbers break into a house. The IoT layer perceives the robbery attempt and immediately informs police authorities. Also, this information gets collected at the fog layer. The FES, having regional knowledge, too classifies this as a potential threat and broadcasts a security message to all the individuals in the locality. Thus all the people in the neighborhood get intimated in time so that they can take proactive security measures. In the third scenario, consider a person who is visiting another city for a routine meeting. However due to maintenance purpose, the intercity highway has been closed. He needs to take a detour through an accident prone secondary route. The situation is worsened due to heavy traffic and presence of dense fog on the route. All this information is readily available to CES in the form of global security base and the profile of the concerned individual. CES concludes that it is unsafe for the person to leave for the meeting and informs him regarding the same.

5   Conclusion

Due to population explosion, it has becoming increasingly challenging for civil authorities to provide security cover to citizens. In order to tackle this enormous challenge, we have proposed a physical security framework that employs the concept of IoT, fog computing and cloud. Security related decisions are taken at various levels or layers depending on their complexity and urgency. The IoT layer interacts with the physical world and gathers knowledge regarding the physical surroundings. It also takes elementary security decisions. The cloud and fog layers take more sophisticated decisions based on complexity of decisions and real time requirements. We have enumerated unique challenges in providing security to citizens using above mentioned technologies. We have also provided a scenario to give insight regarding working of the proposed framework. In the future, we will like to implement and analyze it on actual testbeds and real world scenarios to test its feasibility, practicality and performance. We will also like to tackle the privacy and ethical concerns that come up due to sensitive personal information involved.

References

1. Starner, T.: Project Glass: An Extension of the Self. IEEE Pervasive Computing 12(2), 14–16 (2013)
2. Galaxy Gear, http://www.samsung.com/uk/consumer/mobile-devices/galaxy-gear/galaxy-gear/SM-V7000ZKABTU
3. A new kind of photographic memory, http://getnarrative.com/
4. Nikeplus, http://nikeplus.nike.com/plus
5. Fitbit, http://fitbit.com
6. Schaar, A.K., Ziefle, M.: Smart clothing: Perceived benefits vs. perceived fears. In: 5th IEEE International Conference on Pervasive Computing Technologies for Healthcare, PervasiveHealth (May 2011)
7. Sehgal, V.K., Patrick, A., Rajpoot, L.: A Comparative Study of Cyber Physical Cloud, Cloud of Sensors and Internet of Things: Their Ideology, Similarities and Differences. In: 4th IEEE International Advanced Computing Conference, IACC (February 2014)
8. How Many Internet Connections are in the World? Right. Now, https://blogs.cisco.com/news/cisco-connections-counter/
9. Akyildiz, I.F., Kasimoglu, I.H.: Wireless Sensor and Actor Networks: Research Challenges. Elsevier Ad Hoc Network J. 2, 351–367 (2004)
10. Wlodarczyk, T.W., Rong, C.: On the Sustainability Impacts of Cloud-Enabled Cyber Physical Space. In: 2nd IEEE International Conference on Cloud Computing Technology and Science (2010)
11. Armbrust, M., et al.: A view of cloud computing. Communications of the ACM 53(4), 50–58 (2010)
12. Bonomi, F., et al.: Fog computing and its role in the internet of things. In: Workshop on Mobile cloud Computing, MCC 2012 (2012)
13. Violence against women: an EU-wide survey. Main results report, FRA Publications (2014)
14. Gubbi, J., et al.: Internet of Things (IoT): A vision, architectural elements, and future directions. Future Generation Computer Systems 29(7), 1645–1660 (2013)
15. Xiao, J., Zeng, C., Yu, Z.: The Design of Wild Animals Monitoring System Based on 3G and Internet of Things. TELKOMNIKA Indonesian Journal of Electrical Engineering 11(12), 7762–7768 (2013)
16. Yang, X., Song, W., De, D.: LiveWeb: A Sensorweb Portal for Sensing the World in Real-Time. Tsinghua Science & Technology 16(5), 491–504 (2011)

17. McLean, P.G.: A secure pervasive environment. In: Proceedings of the Australasian Information Security Workshop Conference on ACSW Frontiers 2003, vol. 21, Australian Computer Society, Inc. (2003)

18. Oresko, J.J., Jin, Z., Cheng, J., Huang, S., Sun, Y., Duschl, H., Cheng, A.C.: A Wearable Smartphone-Based Platform for Real-Time Cardiovascular Disease Detection Via Electrocardiogram Processing. IEEE Transactions on Information Technology in Biomedicine 14(3) (May 2010)

19. Stanford, V.: Using Pervasive Computing to Deliver Elder Care. IEEE Pervasive Computing 1(1), 10–13 (2002)

20. Kortuem, G., et al.: Smart objects as building blocks for the internet of things. IEEE Internet Computing 14(1), 44–51 (2010)

21. Hong, K., et al.: Mobile fog: a programming model for large-scale applications on the internet of things. In: Proceedings of the Second ACM SIGCOMM Workshop on Mobile Cloud Computing (August 2013)

22. Zhang, S., Goddard, S.: A software architecture and framework for Web-based distributed Decision Support Systems. Decision Support Systems 43(4), 1133–1150 (2007)

23. Prapakorn, N., Chittayasothorn, S.: An RDF-based distributed expert system. WSEAS Transactions on Computers 8(5), 788–798 (2009)

Classification Mechanism for IoT Devices towards Creating a Security Framework

V.J. Jincy and Sudharsan Sundararajan

Abstract. IoT systems and devices are being used for various applications ranging from households to large industries on a very large scale. Design of complex systems comprising of different IoT devices involves meeting of security requirements for the whole system. Creating a general security framework for such interconnected systems is a challenging task and currently we do not have standard mechanisms for securing such systems. The first step towards developing such a framework would be to build a classification mechanism which can identify the security capabilities or parameters of the different entities comprising an IoT system. In this paper we describe one such mechanism which can take user input to classify the different components of a complex system and thereby determine their capability to support security mechanisms of different degrees. This in turn would enable designers to decide what kind of security protocols they need to adopt to achieve end-to-end security for the whole system.

Keywords: Internet of Things, IoT devices, Properties, Security Properties, Classification, Security Parameters.

1 Introduction

Internet-of-Things (IoT) is the latest paradigm to capture the interest and imagination of not only the scientific and research community but also that of the common man. With the advent of high tech, smart gadgets and the enormous increase in communication bandwidth and availability, the number of novel applications to enhance the quality of life of the common man has increased exponentially within

V.J. Jincy · Sudharsan Sundararajan
Amrita Center for Cyber Security
Amrita Vishwa Vidyapeetham
Kollam, India
e-mail: jincyvalsan87@gmail.com,
sudharsan@am.amrita.edu

© Springer International Publishing Switzerland 2015   265
R. Buyya and S.M. Thampi (eds.), *Intelligent Distributed Computing*,
Advances in Intelligent Systems and Computing 321, DOI: 10.1007/978-3-319-11227-5_23

the last few years. Although the term Internet-of-Things was coined as early as in 1999 when radio-frequency ID (RFID) tags were used to identify and manage products and inventories, the technology has become more pervasive and taken center stage in the recent years. With intelligent systems, automation and remote operations being enabled by the revolution in communication and computing technologies, the importance of the idea of operating in an environment where things interact with human beings and also amongst themselves has become a reality. With such complex systems invading our day-to-day lives the security of information handled and our interactions with such devices takes paramount importance.

Many of our current intelligent systems are a combination of sensors, actuators and networked intelligence. The components comprising these systems can be anything from humans to machines which could be connected and interact amongst themselves. Due to the increasing use of variety of IoT devices in various fields these devices and systems need to have assured secure and safe operation. Currently we do not have specific security mechanisms or standards for different devices and their interoperability. When it comes to the case of large or small IoT systems comprising of various IoT devices the idea of defining a common security framework is yet to be done. The first steps to create a security framework for an IoT system would be to classify the various devices that comprise the system with respect to their capabilities in terms of computing, communication and other important parameters. We have not come across any such direct classification mechanism of this kind in this scenario which would help identify the capabilities of the devices rather than data they handle. In this work we present a different approach to classify the various IoT devices into three distinct classes namely Class A, Class B, Class C with security levels defined as High, Medium and Low respectively. This we take as our first step towards developing a generic security framework for systems that are built by different IoT devices that are interoperable. In Section 2 we describe the existing work that has been done in the field of classification and security of IoT systems. The next section describes about the system design wherein the different properties of the IoT devices which would help in the classification mechanism are listed and discussed. Section 4 describes the shortlisted properties and their usage in the classification mechanism. The section also describes the different IoT systems that were selected as test inputs to the classifier and their classification. We conclude the paper in section 5 stating the use of the classification in creating a security framework for IoT systems.

2   Related Work

Classification becomes important when the concern comes to assigning security levels to IoT systems currently in use and also those which are yet to come. An Ontology based classification [1] was done initially in this field which mainly dealt with two issues that is Heterogeneous Device integration and Composite event detection. In [1] authors have also mentioned about four classes in the first tier which are Places, Nodes, Sensors and Events. The classification was done on wireless sensor networks based on ontology as the name suggests. Next is about the device centric approach for a safer IoT where four different categories were considered by the authors in [2] where devices are supposed to be under risk and these were Hostile Environment, Interference, Misuse, and Internal Failures. The safety concepts were defined staying outside the domains of applications. Context was also a matter of concern there. Context based entities were defined for the first time in [2].The thought of considering the context for the classification came from [2] which was completely novel in the realm of IoT. While considering the security aspect of these devices mainly two types of security [3] was considered namely communication and physical level of security where the communication level [4] was given more importance as the different IoT devices were considered working together in co-ordination. In [4] authors have mentioned about the various issues occurring during the communication of different devices comprising of wireless sensors and actuators which is considered as a reference for analyzing the level of communication security required in each of the listed devices. More description about the security based classification is discussed in the last section. In [5], [6] and [7] the authors have provided a detailed account on the various properties of the IoT devices. They have also discussed about the different scenarios where these are used. In [8] they have discussed in detail about the health status monitoring node from which we could get a detail account about such devices used in the health care. The main concept behind the working of the sensor node is discussed in [9] which have helped a lot in our research work. The network based fire detection system was discussed in [10] which provided us with important information for selecting many of the devices and nodes. In [11] authors have discussed about the devices used for monitoring the environmental parameters. From [12] and [13] we could find some of the devices utilizing the advanced technologies used in industries and make use of their properties for our research. In [15] the authors have done analytical experiments on the working of air traffic control node using formal methods which shows the requirement of a high level of security in the system. Finally in [16] and [17] weka tool (developed by Waikato University in New Zealand) is being discussed which is mainly used for classification, regression, clustering, and association purposes. This tool has been used for the classification in this paper.

3   System Design

Here we describe how we are identifying the various properties of the IoT devices and systems and also discuss about building a classification tree based on the identified properties. This will serve as initial inputs to generate data for the IoT device classification.

3.1 Properties

As a first step we listed out a number of IoT devices and systems and identified their properties which can take on different types of values. Some of the identified properties are discussed below.

Power Constrained or Non Power Constrained. Power constrained systems are mostly working with battery backup and once the battery power is off the system stops working. For example a sensor which is to sense a value every 20 seconds, it senses for 5 seconds and goes off to sleep for 15 seconds thereby conserving power but if this system is compromised then power could be drained by not allowing it to go to sleep. When the power ceases working of the device stops which is a security issue. Non power constrained systems will have a continuous power back up. Values are two

Real Time or Non Real Time. Real Time systems must process information and produce a response within a specified time otherwise results in severe consequences for example Medical devices, Aircraft control and Nuclear Plant management. Non Real time systems process information and produce response but there is no specific time or deadline for that for e.g.: Behaviour Analysis Systems. Values are two.

Closed Loop and Open Loop. Closed loop systems are those which have a feedback mechanism for e.g.: Automated industrial control systems may be used for controlling the amount of water entering into the nuclear reactor at a particular temperature. Open loop systems are those used just for monitoring purposes like weather monitoring. Values are two.

Communication Protocol Used. Devices need to communicate with each other effectively and so different communication protocols are used for communication among these devices e.g.: TCP, UDP, PPP, IGMP, CLNP etc...

Wired and Wireless Systems. Here communication is wired or wireless. It is a physical layer Property. It directly relates to power constrained or non power constrained. Wired examples are RJ45, RS232, Twisted pair, Coaxial cable, Optical fibre Communication. Wireless examples are: Wi-Fi IEEE 802.11, Bluetooth, Infra red, GSM, WiMax IEEE 802.16

Network Size. Based on different application domains size of the network can be the following: Small Medium and large. The application domains can be Smart home/office where the area is very small and number of users is also limited so a small network will do. In case of Smart transportation it includes vast geographical area hence we need a large network size. In case of Smart city we can have a medium network size. Values are: Small, Medium, and Large.

Interoperability. A device can be considered as interoperable if it is compatible in two different networks like for example a device compatible with Bluetooth as well as WiFi.

Reliability. A system is said to be reliable if it has a back up mechanism. (Backing up means copying or archiving data so that the data could be restored once it is lost due to some error or crash) For example in case of sensors deployed in hostile environments if in any case it crashes if it has a backup mechanism then the data sensed so far could be recovered. Hence we can say that the system is reliable but this is not possible then it is an unreliable system

Energy Source. System utilizes energy in different means like it can be Rechargeable battery or Energy harvesting (also known as power harvesting where energy is derived from external sources like solar power, thermal Energy, wind energy, is captured and stored for small, wireless autonomous devices.)So based on the application domain it can be either of them or may be both. In case of Smart home/office it is rechargeable battery, in case of smart city we have both rechargeable battery and energy harvesting.

Processing Speed. It is the measure of cognitive efficiency. The ability to perform a task automatically and speedily in a system. The processing speed of each and every sensor network varies. It can take only a definite value.

Data Management. Data management in case of various systems is basically through Local server or shared server both. In case of smart homes/office a local server does the data management. In case of Smart Transportation shared server does. In Smart Agriculture both local server and shared server does data management. Values are 2: Shared server/Local server

Memory. The storage space required is defined as memory in terms of megabytes, gigabytes etc.... So it can take only a single value. Based on the range of values we can divide it into 3 categories namely low, medium and high for less than 10KB, between 10KB & 1GB and above 1 GB respectively.

Scalability. Depends on the utility of various sensors whether they are used on a large scale, small scale or medium scale. Large scale means usage in an industry in large number, small scale means a limited number of sensors used for a particular purpose may be in a house. Medium scale refers to number of sensors used in places like banks or airports etc...

Bandwidth. The band-width requirement for different systems or devices is different based on their functions. It can take three values .i.e. Low, Medium or High

Size of Sensors. The size of the sensors used in various devices also changes based on the requirements of the device and the application to be performed

Functionality. Functionality means whether it is a critical or noncritical one.

3.2   Classification Tree

Different properties have already been listed in the previous section and based on those properties we built a classification tree in order to identify the important properties among those to be finalized for classification. When the classification tree is built the main node which is the root of the tree is taken to be the important one among the remaining ones (nodes) since the remaining ones can be derived easily from the root. The main node is IoT devices which are further divided into 7 categories and split into sub categories. These categories are size, Type of node, Network layer, System, Scalability, Power and Processing.

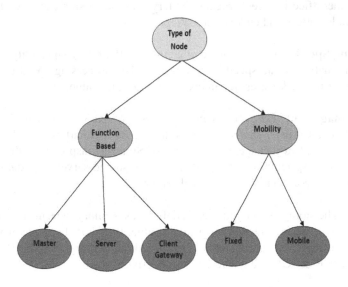

Fig. 1 Classification based on Node Type

Type of node is divided into two namely function based and mobility of the node. This specially relates to the usage scenario of the devices and also the specific nodes acting as master, server and client gateway. The mobility refers to whether the node is movable or not (Fixed or Mobile).

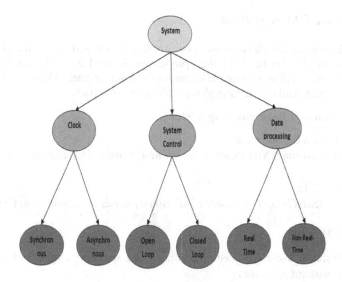

Fig. 2 Classification based on System Parameters

The remaining part is discussed in the next section which is implementation where we have the major task leading to the classification mechanism.

4   Implementation

After the identification of the various properties discussed in the above section the important aspect of the classification is to identify the properties that are important from the standpoint of building a secure system. Hence we have chosen some of the properties from those used to build the classification tree. Also we have chosen a few devices and systems for classification and have discussed their properties and classification

4.1   Finalized Properties

The list of identified properties is as follows

- Connectivity (Wired/Wireless)
- Real-Time (Yes/No)
- Power (Constrained/Non Constrained)
- Device Type (Sensor, actuator, hybrid)
- System Control (Open loop, closed loop)
- Functionality (Critical, Non Critical)

4.2 Device Classification

Here we list out the IoT devices and some applications based on them and the classification of the devices based on their properties. Around 20 devices were selected based on different references and also randomly from internet. These are the list of devices and their brief description along with their properties.

Fire Detection and Extinguishing System

1. Fire Detection Sensor Node
   Wireless, Real time, Non Power Constrained, Sensor, Open Loop, Critical

2. Sprinkler Node
   Wireless, Real-Time, Non Power constrained, Actuator, Open loop, Critical

Industrial Systems

3. Laser Distance Meters: Measure distances and speed in a non contact fashion and also without reflectors
   Wireless, Real Time, Power Constrained, Sensor, Closed loop, Non Critical

4. Thermography Camera module: For accurately measuring or visualizing heat distributions , outputs radiometric images of up to 2048 x 1536 pixel spatial resolution in real time.
   Wireless, Real Time, Power Constrained, Sensor, Closed loop, Non Critical

Home Automation

5. Smart Meters: A **smart meter** is an electrical meter which performs the function of recording the amount of electric energy consumed in intervals of an hour or less and passes on the data to the main utility for monitoring and billing purposes thereby which a two-way communication is established between the meter and the main system and could gather data for remote reporting.
   Wired, Non Real Time, Non Power Constrained, Sensors, Closed loop, Non Critical

6. Motion Detection Sensors in home: A **motion detector** is a device that detects moving objects, particularly people (as they are always in motion). A motion detector is often integrated as a component of a system that automatically performs a task or alerts a user of motion in an area. Motion detectors form a vital component of security, automated lighting control, home control, energy efficiency, and other useful systems.
   Wireless, Real time, Power constrained, Sensor, Open loop, Non critical

7. Smart Bulbs: WiFi enabled, energy efficient LED light bulb that can be controlled using your Smartphone.
   Wired, Non Real Time, Non Power Constrained, Sensors, Closed Loop, Non Critical

Health Care

8. Pacemaker: A pacemaker is a small device placed in the chest or abdomen which helps in controlling abnormal heart rhythms. This device uses electrical pulses to prompt the heart to beat at a normal rate. A pacemaker consists of a battery, then a computerized generator, and also wires with sensors at their tips called electrodes. The generator is powered by the battery, and both of these are surrounded by a thin metal box. The wires connect this generator to the heart.
   Wireless, Real Time, Power Constrained, Sensors, Closed loop, Critical

9. Health Status monitoring node: The integration of various technologies RFID, GPS, GSM and GIS to construct a stray prevention system for the elderly that does not interfere with the elders' daily lives. RFID is the monitoring node.
   Wireless, Real Time, Power Constrained, Sensors (RFID Sensors), Open loop, Critical

Monitoring Nodes

10. Air traffic Control node: It is found that airspace system's capacity and safety are largely dependent on the efficient coordination of air traffic control (ATC) and flight desk personnel. Smart Air Traffic control systems are designed consisting of a Controller, Aircraft, and Network state Controller.
    Wireless, Real Time, Power constrained, Actuator, Closed loop, Critical.

11. Weather monitoring node: Modern weather forecast are result of a computer calculated model which uses weather stations and satellites around the world. It consists of sensors, Timers, multiplexer etc...
    Wireless, Real Time, Power Constrained, Sensor, Closed Loop, Non Critical.

12. Earthquake Detection node: A seismic monitoring and Alarm System have been developed for intelligent structures. It consists of sensor nodes for sensing the earthquake tremors.
    Wireless, Real Time, Power Constrained, Sensor, Closed loop, Critical

13. RFID Smart Conveyor Belt with Confined Detection Range: Radio Frequency Identification (RFID) technology enables detection and recognition of objects associated to univocal identification codes. A typical RFID system comprises an RFID reader linked to one or several antennas that interrogate the tags within its detection range.
    Wired, Real Time, Non power constrained, Sensors (RFID smart conveyor belt), closed loop, Non Critical

14. Land Traffic Control node: Smart Traffic control systems are designed using wireless sensor network which also detects over speeding vehicles.
    Wireless, Real Time, Power Constrained, Sensors, Closed loop, Critical

15. Smart cars sensor node: Modern day cars are equipped with LCD screens, GPS navigation, Sensors telephone and radio controls etc...
    Wired, Real Time, Power Constrained, Sensors, Open loop, Critical

16. Irrigation Systems sensor nodes: Distributed in-field sensor-based irrigation systems offer a potential solution to support site-specific irrigation management that allows producers to maximize their productivity while saving water. Consists of wireless sensor network, software for real time in field sensing.
    Wireless, Real Time, Power Constrained, Sensors, Open Loop, Non Critical

17. Amrita Personal Safety system: Wearable and easy to operate electronic device which helps to trigger communication with family and police when in distress.
    Wireless, Real Time, Power constrained, Actuator, Open Loop, Critical

18. Traffic surveillance camera module: Cameras used for Traffic surveillance especially to point out whether there is any traffic rules violation. CMOS sensors are used.
    Wireless, Real time, Power constrained, Sensor, Closed loop, Non critical.

19. Deep ocean tsunami detection buoys: Deep-ocean tsunami detection buoys are one among the two types of instruments being used by the Bureau of Meteorology (Bureau) to confirm the occurrence of tsunami waves generated by earthquakes undersea. These buoys are meant for observing and recording changes in sea level out in the deep ocean.
    Wireless, Real time, Power constrained, Sensor, Closed Loop, Critical

4.3   Classification Algorithms

There are several classification algorithms which can be used for classifying the different devices we have. Since there were no existing databases for the IoT devices we just created our own database with all devices and their respective properties listed out. Naive Bayes algorithm was found to be appropriate for the classification purpose. Weka tool was used to carry out the whole process of classification where the csv (Comma separated Value) file is provided as input. This file consists of listed properties separated by commas. We get a complete analysis graph and also the confusion matrix from the weka tool from which we can easily make out the different classes as Class A, Class B and Class C. Class A for Critical, Class B for Medium and Class C for Non critical. After the initial classification we switched on to the context based approach in which we included the context in which each device is used as a new property among the existing list of properties. The same procedure was followed as done in previous case mentioned above. Finally we were able to classify the devices to Class A, Class B, and Class C for Critical, Medium and Non critical respectively. We could get the same result as before.

To enhance our findings we considered the security properties as well along with the previous classification and finally obtained a classified set consisting of 3 classes namely Class A, Class B, Class C for High level of security, medium and low level of security respectively. From the obtained result we could infer that our results are correct and if we need to know the security requirement of any IoT device there is just a need to list out properties in csv file and provide as test input. After receiving this test input the weka tool using the Naïve Bayes algorithm compares with the above obtained result and assigns a class for the test input comprising of properties. The output would be displayed in the weka console as one of the three .i.e. Class A, Class B, Class C.As an enhancement to this we can consider the whole system with a wide variety of IoT devices coming under different classes. When the user is able to provide information to a system or tool about the properties of the IoT system to be constructed then using the provided information we can create a csv file which would now be used as a test input and finally the Class of the system could be predicted. This is how we can arrive at the development of a security framework using the existing classification mechanism mentioned in this paper. This security framework would be beneficial in every aspect for the IoT systems yet to be developed.

4.4   Security Framework Creation

The proposed security framework is developed by first grouping the properties of different security protocols like ZigBee, Bluetooth, 6Lowpan, Ethernet and NFC along with the existing security mechanisms in them. The different properties considered were power requirements, Processing power, Net bandwidth, Application, Working Range and network topology. The findings are tabulated and results are obtained after proper analysis of the existing security mechanisms such as the cryptographic techniques and error detection mechanisms being used in those protocols. Based on these properties we carried out the classification as done in the previous section and obtained Class A security suit, Class B suit and Class C suit respectively for Class A, Class B and Class C devices used in creating a new smart system. The created user interface allows the user to enter the details of the devices to be used in creation of new system. Once the details are entered we have a csv file generated with all the properties listed. Once this file is obtained we can manually decide as to which protocol suits it the best based on the previous results and the tabulated data highlighting the security mechanisms used in various protocols mentioned in the beginning. So finally we arrive at a single or may be pairs of protocols compatible with the current system being designed which would provide it the maximum, medium or minimum security based on the context where the system is used. This would be used as the generalized method of assigning security levels to the complex sophisticated smart systems to be designed in the future.

5   Conclusion

The classification mechanism for the IoT devices presented above will help to identify the different security capabilities of the devices by assigning them a particular class and at the same time help the system designers to evolve a security mechanism for the whole system taking into consideration the different classes of devices that need to interact in the system. This will help to act as a foundation for building a generic security framework guideline wherein a system designer can choose from a range of security protocols available for different classes of devices which he can integrate to achieve the required security level for the whole system.

References

1. Danieletto, M., Bui, N., Zorzi, M.: An Ontology-based Framework for Autonomic Classification. In: 2011 IEEE International Conference on Internet of Things Communications Workshops (ICC), pp. 5–9 (2011)
2. Chen, C., Helal, S.: A Device-Centric Approach to a Safer Internet of Things. In: NoME-IoT 2011 Proceedings of the 2011 International Workshop on Networking and Object Memories for the Internet of Things. ACM, New York (2011)
3. Weingart, S.H.: Physical Security Devices for Computer Sub systems: A Survey of Attacks and Defenses. In: CHES 2008 Version (2008)
4. Slijepcevic, S., Potkonjak, M., Tsiatsis, V., Zimbeck, S., Srivastava, M.B.: On Communication Security in Wireless Ad-Hoc Networks. In: Proceedings of the Eleventh IEEE International Workshops on Enabling Technologies: Infrastructure for Collaborative Enterprises (WETICE 2002) (2002)
5. Yang, D.-L., Liu, F., Liang, Y.-D.: A Survey of the Internet of Things. In: The International Conference on E-Business Intelligence (2010)
6. Gubbia, J., Buyyab, R., Marusic, S., Palaniswami, M.: Internet of Things (IoT): A vision, architectural elements, and future directions
7. Agrawal, S., Vieira, D.: A survey on Internet of Things
8. Barger, T.S., Brown, D.E., Alwan, M.: IEEE Health-Status Monitoring Through Analysis of Behavioral Patterns. IEEE Transactions On Systems, Man, And Cybernetics (January 2005)
9. Chong, C.-Y., Kumar, S.P.: Sensor Networks: Evolution, Opportunities, and Challenges. Proceedings of the IEEE (August 2003)
10. Lee, K.C., Lee, H.-H.: Network-based Fire-Detection System via Controller Area Network for Smart Home Automation. IEEE Transactions on Consumer Electronics (November 2004)
11. Haefke, M., Mukhopadhyay, S.C., Ewald, H.: A Zigbee Based Smart Sensing Platform for Monitoring Environmental Parameters. In: 2011 IEEE Instrumentation and Measurement Technology Conference, I2MTC (2011)
12. Official website of Jenoptik http://jenoptik.com (accessed on November 2013)
13. Official Website of Beecham Research, http://beechamresearch.com/article.aspx?title=Research (acesses on November 2013)

14. Wikipedia on Internet_of_Things: wikipedia.org/wiki/Internet of Things
15. Jamal, M., Zafar, N.A.: Requirements Analysis of Air Traffic Control System Using Formal Methods
16. Official Website of waikato university, http://cs.waikato.ac.nz/ml/ (accessed on November 2013)
17. Frank, E., Trigg, L., Hall, M., Holmes, G., Kirkby, R., Pfahringer, B., Witten, I.H.: Weka: A Machine learning Workbench for data mining

Data Owner Centric Approach to Ensure Data Protection in Cloud Environment

Kanupriya Dhawan and Meenakshi Sharma

Abstract. Last few years the latest trend of computing has been cloud computing. Cloud has brought remarkable advancement in individual as well as IT sector. But still many organisations are lagging behind in using cloud services. Major issue that affect them is related to data protection at cloud and afraid of sensitive data leakage by intruders. To solve this problem a data protection model has been designed so that data owner feels free to use cloud services. Proposed model is highly secure and data is under control of data owner itself. Re-encryption, HMAC and Identity based user authentication techniques are used in this model to make it more effective and attractive to use this model in real world.

Keywords: Cloud Computing, Re-encryption, Data Security, HMAC, Data Privacy.

1   Introduction

Cloud computing has become biggest recent emerging trend from past few years that have gather huge curiosity in world of computing. Cloud technology has brought remarkable advancement as clients do not have to upgrade their systems for work. Basically Cloud is pay-as-you-use service to access hardware/software or storage or full virtual machine available 24*7.Initially, organization store sensitive data internally protected with different security means but now days many organisation started storing their data towards cloud. Still some organisation lags behind in adoption of cloud due to security. Data security is most prominent issue of cloud computing. When the users upload their data to cloud then they have no direct hold over data. They only have to trust the cloud service provider for data security. When these providers are untrusted then data is unsafe. Data owner

Kanupriya Dhawan · Meenakshi Sharma
SSCET Badhani, Pathankot
e-mail: unak.211990@gmail.com

© Springer International Publishing Switzerland 2015   279
R. Buyya and S.M. Thampi (eds.), *Intelligent Distributed Computing*,
Advances in Intelligent Systems and Computing 321, DOI: 10.1007/978-3-319-11227-5_24

cannot trust anyone while outsourcing its data. Its main role is to protect data by themselves.

On the basis of main responsibility of data owner towards data security an approach of data owner centric protection has been designed in form of model. Data is protected against third party, network intruder and cloud service provider. Proxy Re-encryption, HMAC and identity based user authentication technique has been used in this model. Proposed model give data owner control over its data store at cloud and make them feel free without any fear of losing their data at cloud. The rest of paper has been assembled as Section 2 represents the related work done in the field of data protection and privacy. Section 3 describes proposed data owner centric protection model. Section 4 we conclude our work.

2   Related Work

There are numerous work carried in the field of data protection at cloud. Many models, schemes and techniques are proposed for data security. Hwang et.al[1] proposed business model in which encryption/decryption service and storage as a service of user data were separated i.e. they were not provided by single operator. After encryption/decryption performed system should delete all the data.

Varalakshmi et.al[3] proposed system consists of three entities cloud broker, client and cloud storage. Broker handles encryption, hash key, decryption and local database management. According to cloud space available the client files are partitioned into segment and hash values of segments has been generated. When the client needs its file it sends request to broker then broker download the file, partition the file into segments and then calculate the hash values. For checking the data integrity hash values before uploading to the after downloading are matched. If this matches data is un-tampered. Mohamed et.al[4] performed randomness testing on various eight encryption technique namely RC4, RC6, MARS, AES, DES, 3DES, Two-Fish and Blowfish.

Xu et.al[5] propose a dynamic user revocation and key refreshing scheme based on cipher policy attribute based encryption technique. In this technique user can be removed anytime without changing keys and also refresh keys without re-encrypting data. Huang et.al[6] proposed scheme that consists of four entities – SSManager, SSGuard, SSCoffer and user. SSGuard do encryption before uploading and uploaded files store at SSCoffer. File encryption key are encrypted by user public key and store at SSManger. For decryption of file QR code is used. User shows QR code to SSGuard to decrypt files. Sur et.al[7] proposed a model in which certificate based Proxy re-encryption scheme is followed before uploading data to cloud. Mowbray et.al gives general overview of protecting data in cloud and describes various approaches to handle this protection. Our proposed model provides data protection at cloud and data security from internal as well as external threats. This model is organised in way that data remains private throughout transits and at rest.

3   Proposed Model

Proposed model is highly based on data owner centric approach i.e. data owner is responsible for data protection in cloud and it cannot trust anyone like cloud service provider or third party for data security. The outcome of model is extremely productive in terms of security of data is concerns i.e. data remains private even if it is outsourced. To maintain data privacy re-encryption is performed with the help of third party and for data integrity Hash Based message authentication code is generated on encrypted data. The model is divided into two phases and consists of data owner, third party, cloud service provider and user.

The phases are:

- Phase I (Uploading or Data Storage)
- Phase II (Downloading or Data Retrieval)

3.1   Phase I- Uploading

3.1.1   Key Generation and Distribution

In this security model data owner generate the Keys and divide the keys into key pieces. Owner keeps his piece with it for encryption. Distribute other piece of key to third party for re-encryption and also store key pieces for corresponding to user id for later use.

3.1.2   Re-Encryption and Indexing

For maintaining data privacy, data is encrypted and then uploaded to cloud. In this model owner perform encryption on data with its key piece and give encrypted data to third party. Third party re-encrypt the data with its key piece and then upload to cloud.

As it is very complicated to search on encrypted data so indexing of data has been made by owner. Indexing is way to retrieve data faster. As we do not want our data disclosure to cloud provider or network attacker or third party so index is encrypted by owner first with its key piece. Further pass to third party to re-encrypt it before uploading to cloud.

3.1.3   HMAC for Data Integrity

Although data is in encrypted format but there is fear of data being tampered during transit or on storage of data. To resolve this fear of data tampering hash-based message authentication code (HMAC) is calculated after data encryption. Basically, HMAC is process to use cryptographic hash function for message in the form of data authentication. It is encrypted by owner then re-encrypted by third party and uploaded along with data to cloud.

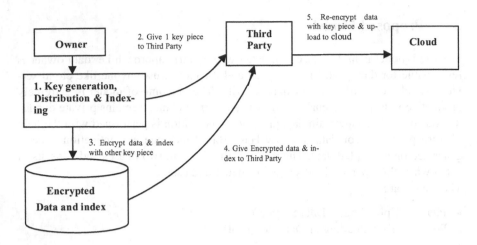

Fig. 1 Re-encryption and Indexing

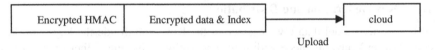

Upload

Fig. 2 HMAC Generation and Uploaded to cloud

3.2 Phase II (Downloading or Data Retrieval)

3.2.1 Identity Based Data Access and User authentication

In the proposed model data is protected against unauthorized user so identity based access has been provided to data. For user authentication owner has already created database related to user login detail and identity base access at cloud. On successful login to cloud by user, user get identity based access to encrypted data. User takes its data to third party to decrypt it. Firstly third party verifies the user then decrypts the data with its key piece. Still data is in encrypted format now passed to data owner by third party for decryption with respective user id. Owner verifies user id, as it is identity base access the owner take respective user key piece corresponding to user id and decrypt index. Passes user key piece to user so that it can decrypt its data according to its identity access.

3.2.2 HMAC Generation

After decryption by third party, the data is still in encrypted format and passed to owner. Then data owner generate HMAC of data. If HMAC before uploading the data to cloud equals HMAC after downloading data then data integrity holds else data has been tampered by some intruder.

HMAC(uploading) = HMAC(Downloading) ⟶ Data is un-tampered

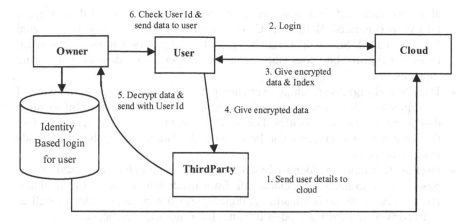

Fig. 3 Identity Based Data Access and User authentication

4 Security Analysis

Proposed model is secure from all threats prevailing and harming the data throughout the phases. It provides security when data at rest as well as during transmission. Data is protected against authorized access by third party or cloud provider itself or network intruder. This model is designed in way to protect data from every aspect. Security analysis of model has been performed on basis of some parameters and comparative analysis is performed between existed model and proposed model. The parameters are:

- **Data Confidentiality:** In the proposed model for data secrecy re-encryption technique has been designed. The data is encrypted by two authorities i.e. data owner then by third party and after re-encryption has been performed data has been uploaded to cloud. In case third party is untrusted then also data is secure as data goes to third party after encryption for re-encryption. All the models in comparative analysis provide data confidentiality.
- **Data Privacy:** The data remains private throughout the two phases in the model. There has been no disclosure of data without verification. Untrusted parties are taken into consideration while designing model and data remains private against them. Hwang et.al [1] violates data privacy as upload of plain text data by user on cloud. During transit data remains endangered. Varalakshmi et.al [3] proposed model which take help of broker for data encryption and if broker is untrusted then data privacy is violated.
- **Overhead:** As it is owner centric approach so all the overhead will on data owner but this model has been proposed such that data owner overhead should be less. For encryption and user authentication owner helping hand is third party and cloud provider itself. But the data is protected against them too. Xu et.al[5] propose a dynamic user revocation and key refreshing scheme and this scheme is data owner centric so all the overhead over data owner.
- **Data Integrity:** To check that data is safe during transit as well at rest Hash Based Message Authentication Code is generated. HMAC is also encrypted

along the data and then uploaded to cloud. When file is needed then again HMAC is generated. Both HMAC are matched for data integrity. Hwang et.al [1], Xu et.al[5], Sur et.al[7] and Huang et.al[6] all proposes model focus mainly on encryption but does not check for tampering of data or not during transmission.

- **Data Searching:** As searching over encrypted data is very tough so indexing of encrypted data has been made. Index is also encrypted so that no information disclosure to untrusted parities. During Decryption Index is decrypted first so that Search over encryption can be performed. Data searching has been made fast due to indexing.
- **User Authentication:** Identity based user authentication has been used in proposed model so that data remains safe from inside attacks as well as unauthorized access. User authentication is done by cloud service provider as well as third party. Other exiting models does not focus not user verification.
- **Availability:** According to SLA certification of cloud service provider the data will be available 24*7. The data can be access anytime from anywhere.
- **Inside Attack:** It refers to not loyal employees/users of an organisation. To protect data from this type of users, identity based user access have been implemented in the model. User cannot have access to full data. According to their identity in an organization can access data. No focus on Users verification in exiting models but Xu et.al [5] proposes technique based user revocation and key refreshing i.e. user can be change anytime without changing of key.
- **External Attack:** External Attack can be from untrusted parties like third party or cloud service provider or network intruder. The data is safe as it is in encrypted format throughout different the phases of model. Hwang et.al [1] proposes model in which cloud service provider help owner to protect data and if cloud service provider is untrusted then data endangered. Varalakshmi et.al [3] also proposes in which cloud broker plays vital. So both the model are prone to external attack.

Table 1 Comparative Analysis

Parameters	Hwang et.al [1]	Varalakshmi et.al[3]	Xu et.al[5]	Sur et.al[7]	Huang et.al[6]	Proposed model
Data Confidentiality	Yes	Yes	Yes	Yes	Yes	Yes
Proxy re-encryption	No	No	No	Yes	Yes	Yes
User Authentication	No	No	No	No	No	Yes
Overhead	Less	Less	More	Less	Less	Less
Data availability	Yes	Yes	Yes	Yes	Yes	Yes
Data Integrity	No	Yes	No	No	No	Yes
Data Privacy	No	No	Yes	Yes	Yes	Yes
Inside Attack	Not secure	Not secure	secure	Not secure	Not secure	Secure
External Attack	Not secure	Not secure	secure	secure	secure	Secure

5   Conclusion and Future Scope

Recent Trends of computing shows cloud computing has made incredible development from past few years. Still there are certain issues which are obstacles in the growth of cloud computing. Most prominent issue prevailing now days is data security. To overcome this issue model has been proposed which is secure enough to use in cloud for storing data. Proposed model is divided into two phases i.e. data storage and data retrieval. During two phases data remains private and not disclose to unauthorized access. This model is data owner centric with least overhead and highly secure to adopt in real life while storing and retrieving data from cloud. Proposed model will be applied to realistic project for enhancing effectiveness of our model.

References

1. Hwang, J.-J., Chuang, H.-K., Hsu, Y.-C., Wu, C.-H.: A Business Model for Cloud Computing Based on a Separate Encryption and Decryption Service, National Science Council of Taiwan Government
2. Fan, C.-I., Huang, S.-Y.: Controllable Privacy Preserving Search Based on Symmetric Predicate Encryption in Cloud Storage. In: International Conference on Cyber-Enabled Distributed Computing and Knowledge Discovery. IEEE (2011)
3. Varalakshmi, P., Deventhiran, H.: Integrity Checking for Cloud Environment Using Encryption Algorithm. IEEE (2012)
4. Mohamed, E.M., EI-Etriby, S.: Randomness Testing of Modem Encryption Techniques in Cloud Environment. In: 8th International Conference on Informatics and Systems (2012)
5. Xu, Z., Martin, K.M.: Dynamic User Revocation and Key Refreshing for Attribute-Based Encryption in Cloud Storage. In: International Conference on Trust, Security and Privacy in Computing and Communications. IEEE (2012)
6. Huang, K.-Y., Luo, G.-H., Yuan, S.-M.: SSTreasury+: A Secure and Elastic Cloud Data Encryption System. In: International Conference on Genetic and Evolutionary Computing. IEEE (2012)
7. Sur, C., Park, Y., Shin, S.U., Seo, C., Rhee, K.H.: Certificate-Based Proxy Re-Encryption for Public Cloud Storage. In: International Conference on Innovative Mobile and Internet Services in Ubiquitous Computing. IEEE (2013)
8. Zhu, S., Yang, X., Wu, X.: Secure Cloud File System with Attribute based Encryption. In: International Conference on Intelligent Networking and Collaborative Systems. IEEE (2013)
9. Mowbray, M., Pearson, S.: Protecting Personal Information in Cloud Computing. Springer-Verlag (2012)
10. Liu, J., Wan, Z., Gu, M.: Hierarchical Attribute-Set Based Encryption for Scalable. In: Flexible and Fine-Grained Access Control in Cloud Computing. Springer-Verlag (2011)
11. Sood, S.K.: A Highly Secure Hybrid Security model for Data Security at Cloud. Submitted to Security and Communication Networks. John Wiley and Sons, Interscience (2012); Special Issue on Trust and Security in Cloud Computing

12. Sood, S.K.: A Combined Approach to Ensure Data Security in Cloud Computing. Submitted to Journal of Network and Computer Applications (2012)
13. Buyya, R., Vecchiola, C., Selvi, S.T.: Mastering Cloud Computing Foundations and Applications Programming. Morgan Kaufmann, USA
14. Cheng, F.: Security Attack Safe Mobile and Cloud-based One-time Password Tokens Using Rubbing Encryption Algorithm. Springer Science+Business Media (2011)
15. Modi, C., Patel, D., Borisaniya, B., Patel, A., Rajara, M.: A survey on security issues and solutions at different layers of Cloud computing. Springer (2012)

Predictive Rule Discovery for Network Intrusion Detection

Kanubhai Patel and Bharat Buddhadev

Abstract. Good number of rule-based intrusion detection systems (IDS) is widely available to identify computer network attacks. The problem with these IDS is requirement of domain experts to constitute rules manually. Rules are not learned by these IDSs automatically. This paper presents novel technique for predictive rule discovery using genetic algorithm (GA) approach to create rule base for IDS. The motivation for applying GA to rule discovery is that it is robust and adaptive search technique in nature that perform a universal search in the solution domain. KDD Cup 99 training and testing datasets were used to generate and test rules. We have obtained 98.7% detection rates during testing to detect various types of attacks.

1   Introduction

Computer network attacks are becoming more complex, sophisticated, and dangerous. Due to this it is required to improve the functionality and quality of security systems; such as intrusion detection systems (IDS) continually [1]. IDS provide security for computer network systems by distinguishing between normal and malicious activity. IDS employ techniques for modeling and recognizing intrusive behaviour in a computer system [2]. There are two main detection methods, (i) misuse detection, and (ii) anomaly detection [2] [3]. These terms are also known as knowledge based and behaviour based intrusion detection [4] respectively.

Kanubhai Patel
CMPICA, Charotar University of Science & Technology, India
e-mail: kkpatel7@gmail.com

Bharat Buddhadev
SS College of Engineering, Bhavnagar, India
e-mail: bvbld@yahoo.com

© Springer International Publishing Switzerland 2015
R. Buyya and S.M. Thampi (eds.), *Intelligent Distributed Computing*,
Advances in Intelligent Systems and Computing 321, DOI: 10.1007/978-3-319-11227-5_25

The misuse detection method attempts to encode knowledge of known intrusions, typically as rules, and use this to screen events [5] [6]. The anomaly detection method try to learn the features of event patterns of normal behaviour and detect intrusions by observing patterns that deviate from established norms [7].

A good number of soft computing based methods and solutions have been proposed for network intrusions detection [8] [9] [10] [11] [12] [13] [14] [15] [16] [17] [18] [19]. As per Gong, Zulkernine, & Abolmaesumi [11],

"Soft computing is a collection of techniques that take advantage of the uncertainty, forbearance for inaccuracy, partial truth, and estimate to achieve robustness with inexpensive solution."

The main techniques of soft computing are ant colony optimization, particle swarm intelligence, fuzzy logic, artificial neural networks, probabilistic reasoning, and genetic algorithm (GA) [19]. Soft computing techniques are often used in combining with rule-based expert systems, when used for network intrusion detection. These rule-based expert systems acquire expert knowledge [20] [14] [18] [17]. The knowledge is described as a set of IF-THEN rules.

A Good number of rule-based intrusion detection systems (IDS) is widely available to identify attacks. The problem with these IDS is requirement of domain experts to constitute rules manually. Rules are not learned by these IDS automatically. This paper presents novel technique for rule discovery using genetic algorithm (GA) approach to create rule base for IDS. The motivation for applying GA to rule discovery is that they are robust and adaptive search techniques in nature that perform a universal search in the solution domain. KDD Cup 99 [21] training and testing datasets were used to generate and test rules. As per Gong, Zulkernine, & Abolmaesumi [11], although various soft computing based techniques and solutions having been proposed, potential of these techniques for intrusion detection are still not fully explored and utilized.

We have presented review of related works in Section 2. A brief overview of GA is described in Section 3. Section 4 presents our proposed GA based rule discovery approach. Section 5 describes experimental results along with dataset description. In last, Section 6 describes conclusion and future works.

2   Related Works

Denning [7] proposed an intrusion detection model in 1987 first time. After that a good number of research efforts has been carried out on the effective detection of intrusion. Besides other approaches, a good number of soft computing based approaches has been proposed to detect network intrusions. A good number of researchers have used GA for deriving classification rules [11] [22] [15] [23] [24] and performing various functions of IDS [8] [10] [11] [12] [13].

Gong, Zulkernine, and Abolmaesumi [11] have presented an implementation of GA based approach to derive a set of classification rules. Support-confidence framework to judge fitness function was used by them. Li [22] has presented a technique to detect anomalous network intrusion using GA. He considered both quantitative and categorical features to derive classification rules.

Chittur [15] has used GA for anomaly detection, while Goyal and Kumar [25] proposed a GA based algorithm to derive classification rules for all types of smurf attack. Ashfaq, Farooq, and Karim [23] have proposed a Cost-Sensitive misuse detection technique for efficient rule generation using GA. GA was used to optimize the features or parameters requirements of some main tasks in which different computational intelligence techniques are used to derive rules [20] [14]. Mischiatti and Neri [26] have used REGAL, a distributed GA, for evolving rules. They have compared the performance of REGAL against RIPPER.

Some of researchers have combined fuzzy logic with GA to discover rules for intrusion detection [27] [28] [29] [30] [31]. Using a fuzzy rule-based genetic classifier, Abadeh and Habibi [32] and Tansel et al [33] have proposed techniques based on iterative rule learning. Wang et al. [34] have used the evolutionary algorithm to evolve neural networks for intrusion detection.

Marin, Ragsdale, and Surdu [35] have developed a hybrid technique that combines expert systems, clustering techniques, GA, and Linear Vector Quantization (LVQ) for generating user profiles. Gomez et al [36] have proposed a linear representation scheme to derive fuzzy rules using the theory of complete binary tree structure. Hofmann et al [37] have presented the evolutionary learning of radial basis function (RBF) networks for network intrusion detection. Martin volkerbutz [38] has investigated rule-based evolutionary online learning classifier systems (LCSs). Abdullah et al [12] have used GA to detect anomalous network behaviors based on information theory.

Lin and Wang [8] have proposed a new genetic clustering algorithm for intrusion detection. They got the optimum number of clusters and high performance rates by this new algorithm. S. Ganapathy, K. Kulothungan, P. Yogesh, A. Kannan [39] have proposed a Novel Weighted Fuzzy C-Means clustering method based on Immune GA (IGA-NWFCM) to find the proper cluster structures from a dataset used for intrusion detection. Wu and Banzhaf [10] have proposed a multilevel genetic programming approach based on their computational multilevel selection framework [9]. Amin and Radu [40] have proposed cooperative intrusion detection functions that take into account multiple objectives simultaneously.

Sen and Clark [41] have explored the use of genetic programming and grammatical evolution to evolve intrusion detection programs for mobile ad hoc networks (MANETs). Balajinath and Raghavan [42] have applied a GA to perform intrusion detection based on UNIX commands. Hoque, Mukit, and Bikas [13] have presented network based IDS using GA. Although a good number of researchers have used GA based techniques, still there is scope of improvement in using the GA based technique for intrusion detection.

3   A Brief Overview of Genetic Algorithm

As per Miller and Shaw [43],

"Genetic Algorithm (GA) is an efficient searching technique used in computing to locate precise or estimated solutions to optimization and search problems based on the principles of genetics and natural selection."

GA has been successfully applied in many optimization and machine learning problems. A GA allows a population to evolve to a state that maximizes the fitness. This GA based method was developed by John Holland [44]. It was popularized by David Goldberg [45].

The algorithm starts by randomly generating a large population of candidate programs. GA works in an iterative manner by generating new individuals or populations from old ones. Every individual is the encoded in binary, real, or even rule etc.

An evaluation function calculates a fitness value of every individual indicating its fitness for the problem. Some type of fitness measure to evaluate the performance of each individual in a population is used. A large number of iterations is then performed that low performing programs are replaced by genetic recombination of high performing programs. That is, a program with a low fitness measure is deleted and does not survive for the next computer iteration [46]. Researchers have tried to integrate it with IDS.

Algorithm 1. Generic Pseudo-code for GA

```
1  GA_Algorithm
2  {
3     Generate initial population of
       individuals randomly i.e. Pop[0]
4     Gen = 0
5     Compute the fitness of each
       individuals (in Pop[0])
6     while not done do //test for
       termination criterion
7     {
8        Gen = Gen + 1
9        Select individuals based on
          fitness (i.e. Pop[Gen] from
          Pop[Gen - 1])
10       Apply Crossover and Mutation
          to selected individuals,
          creating new population/individuals
11       Compute fitness of each of the
          new individuals
12       Update the current population
          //new individuals replace old
          individuals
13    } //while
14 } //end of GA_Algorithm
```

4   The Proposed GA based Rule Discovery Technique

We have tried to derive predictive rules using GA based approach. This section presents a GA based approach that evolves a population of individuals. Here each individual represents a predictive rule. We encoded the antecedent (**IF** part) of the predictive rule in each individual. The consequent (**THEN** part) of the predictive rule is not encoded in the individual genome. We will fix it for a given GA run. So, all the individuals are considered for rules with the same consequent in in each GA run. The other information of our GA approach is as given below.

4.1   Individual Representation

In the proposed approach, the genome of an individual comprises of a union of conditions representing antecedent of a rule. As shown in Figure-1, an individual is represented as a set of k conditions. Here k is the number of attributes. A rule condition is represented by a gene in the form *Ak Opk Vkj*, [47] where:

- *Ak* denotes the k-th predictor attribute;
- *Opk* denotes the relative operators, e.g. "<=" or ">" for continuous attributes, and "=" for categorical attributes used in the k-th conditions;
- *Vkj* denotes the *j*-th value of the *Ak* domain.

A1 Op1 V1j ;	A2 Op2 V2j;	.....	Ak Opk Vkj

Fig. 1 Structure of Genome of an individual [47]

Position of attribute is maintained in the genome. The value of gene is assigned as #, if that attribute is not part of the rule antecedent. This indicates that the attribute is not a part of the rule antecedent. An example of a rule (/individual) (whose genotype has the structure is as shown in Figure-1) is shown in Figure-2.

protocol_type = tcp	#	src_bytes = 0	....

Fig. 2 Example of Genotype

In KDD Cup 99 dataset, data being mined has 41 attributes. In Figure-2, first and third attributes have values, so two conditions are shown in the genotype. The above genotype is interpreted into the following rule antecedent.

IF *Protocol_type = tcp and src_bytes = 0.*

The predicted class is specified by the consequent part of the rule. This is not represented in the genome.

4.2 Genetic Operators

Genetic operators (i.e. selection, crossover, and mutation) are used in GA to keep the genetic diversity by bringing in new genetic materials and by manipulating and/or recombining the genetic material of individual rules.

Selection: In our proposed approach, tournament selection is used to select the individuals with tournament size of 5. We use three-point crossover with 95% crossover probability and 1% mutation probability. Elitism is used with an elitist factor of 2, means the best two individuals of each generation are passed unaltered into the next generation.

Crossover: Two parents are selected and perform swapping of two genes which are randomly selected. Note that inter-genes crossover is performed, not intra-gene crossover. So rule conditions are swapped between individuals. New rule conditions are not generated.

Mutation: At a time, mutation is carried out on a single individual by transforming randomly the value of an attribute into different value that fit in to the range of that attribute. By this way, it helps us to get the universal best solution of the problem.

4.3 Fitness Function

The fitness function is a single figure or value to decide quantitatively that an individual is fit to solve the target problem. It is mainly used to determine good individuals to use for reproduce. An individual with high fitness value has the higher probability of selecting that to use for reproduction, crossover, and mutation operations.

The classification rule (R) is in the form "**IF** A **THEN** C" where A is antecedent part and C is the consequent part. We measured confidence, coverage and simplicity of the rule to calculate the fitness value of the rule.

The confidence (Cf) of the rule R can be computed as:

$$Cf(R) = | nA * nC | / nA. , \qquad (1)$$

here, nA is the number of instances that satisfy the antecedent part and nC is the number of instances that satisfy the consequent part.

The coverage (Cv) of the rule R can be computed as:

$$Cv(R) = | nA * nC | / nC. , \qquad (2)$$

The coverage of the rule shows that how rule is complete in solving the problem.

When a rule has at most L number of conditions, the comprehensibility (or simplicity) of the rule R can be computed as:

$$Cp\ (R) = L - len\ /\ L - 1, \hspace{3cm} (3)$$

here, len is the length (i.e. number of condition) of the rule R.

When a rule contains more attributes then it is considered as complex. The fitness of a rule is calculated as the weighted mean of confidence, coverage and comprehensibility as under.

$$Fc\ (R) = wa * Cf\ (R) + wb * Cv\ (R) + wc * Cp\ (R). , \hspace{2cm} (4)$$

Where wa, wb, and wc are the constants or weights for various factors. We have tested our proposed technique using the weight or values of wa=0.6, wb=0.3 and wc=0.1 according to our assessment of relative importance of confidence, coverage and comprehensibility. The value of Fc (R) is normalized in the range of 0 and 1.

5 Experimental Results

5.1 Dataset Description

The KDD Cup 99 dataset is the most popular benchmark dataset that is used in the intrusion detection research field. There are 41 features which described each TCP connection. These records also have a label which specifies whether the connection status is normal or intrusive [21].

There are 38 numeric features and 3 symbolic features, falling into the following four categories:

- Basic features: 9 basic features were used to describe each individual TCP connection.
- Content features: 13 domain knowledge related features were used to indicate suspicious behavior having no sequential patterns in the network traffic.
- Time-based traffic features: 9 features were used to summarize the connections in the past 2 seconds that had the same destination host or the same service as the current connection.
- Host-based traffic features: 10 features were selected.

The training set contains 4,940,000 data instances, covering normal network traffic and 24 attacks. The test set contains 311,029 data instances with a total of 38 attacks, 14 of which do not appear in the training set. Since the training set is prohibitively large, another training set which contains 10% of the data is frequently used.

Attacks fall into mainly four categories:

- DOS: denial-of-service, e.g. land, pod, back, teardrop, apache2, smurf, neptune, dosnuke, tcpreset, syslogd, crashiis, arppoison, mailbomb, selfping, processtable, udpstorm, warezclient;

- R2L: unauthorized access from a remote machine, e.g. guessing password, sendmail, imap, dict, netcat, ncftp, xlock, xsnoop, sshtrojan, framespoof, ppmacro, guest, netbus, snmpget, ftpwrite, httptunnel, phf, named;
- U2R: unauthorized access to local super user (root) privileges, e.g.,sechole, xterm, eject, ntfsdos, nukepw, secret, perl, ps, yaga, fdformat, ppmacroffbconfig, casesen, loadmodule, sqlattack;
- Probing: surveillance and other probing, e.g., portsweep, ipsweep, lsdomain, ntinfoscan, mscan, illegal-sniffer, queso, satan.

5.2   Implementation of the System

The proposed GA based technique is implemented in software using Java programming language. We have created classes for individual, population, rule_evaluator (for fitness calculation), and breeder (for crossover and mutation). Individual class describes chromosome representation and fitness value. Population class initializes the population. Rule_evaluator class is for to calculate the fitness according to the fitness function described in Section 4-3. Breeder class is for performing crossover and mutation to generate new population.

In first phase, application use KDD Cup 99 dataset to generate rule base. The system was trained by setting the parameters as below:

- No of generations: 500
- Cross over: one-point cross over
- Probability of crossover: 0.7
- Mutation rate: 0.01

In second phase, we test the system using this rule base and the KDD Cup 99 testing dataset.

5.3   Results

In the testing phase, we have obtained 98.85% and 99.19% detection rates for first and second experiments for normal samples. While for various attacks we have obtained detection rates as shown in Table-1 and in Fig-3.

Table 1 Detection rates for various attacks

Sr.	Attacks	Detection rate (%) in experiment	
		I	II
1	Smurf	98.23	98.78
2	Neptune	99.56	99.42
3	Back	98.45	99.12

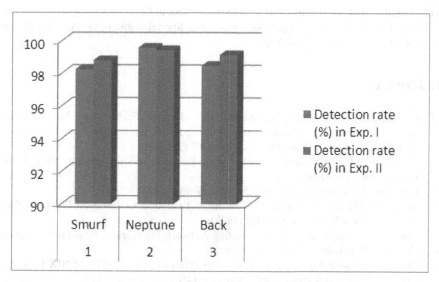

Fig. 3 Detection rates for various attacks in Exp-I and Exp-II

We have obtained 98.7% overall detection rate during testing phase for various type of attacks. Following are some of the examples of rules generated by the system after training. These rules are used to test the system to detect various attacks.

IF protocol_type=tcp and service=http THEN back

IFprotocol_type=tcp and service=http and logged_in=1 and dst_host_count =255 and dst_host_same=0 THEN back

IF protocol_type=tcp and service=private THEN Neptune

6   Conclusions and Future works

In this work, we attempt to explore the use of GA for discovering predictive, simple, and complete classification rules. The uniqueness of our proposed GA based approach is in flexible individual representation, suitable genetic operators, and effective fitness function. The proposed approach is tested on KDD Cup 99 datasets and the classification rule sets discovered based on three measures, i.e. accuracy, converge and comprehensibility. Our approach gives higher predictive accuracy for the DoS attack as compared other type of attacks.

As future work, we would like to compare the proposed GA based approach with C4.5 and DTGA algorithms with respect to the above three measures of the derived rules.

Acknowledgments. We are thankful to the management of Charotar University of Science & Technology for providing support for the research. Special thanks to Dr. Ajay Parikh and Dr. S K Vij for their support and help.

References

1. Gomez, J., Gill, C., Banos, R., Marquez, A., Montoya, F., Montoya, M.: A Pareto-based multi-objective evolutionary algorithm for automatic rule gen-eration in network intrusion detection systems. Soft Computing 17 (2013)
2. Engen, V.: Machine Learning For Network Based Intrusion Detection, PhD thesis, Bournemouth University (2010)
3. Gollmann, D.: Computer Security, 2nd edn. Wiley (2006)
4. Debar, H., Dacier, M., Wesp, A.: Towards a Taxonomy of Intrusion Detection Systems. Computer Networks, 805–822 (1999)
5. Paxson, V.: Bro: A System for Detecting Network Intruders in Real-Time. Elsevier Computer Networks, 2435–2463 (1999)
6. Roesch, M.: Snort: Lightweight intrusion detection for networks. In: SENIX Large Installation System Administration Conference, LISA, pp. 229–238 (1999)
7. Denning, D.: An intrusion-detection model. IEEE Transactions on Software Engineering 13(2), 222–232 (1987)
8. Lin, C., Wang, M.: Genetic-clustering algorithm for intrusion detection sys-tem. International Journal of Information Computer Security 2(2), 218–234 (2008)
9. Wu, S., Banzhaf, W.: A hierarchical cooperative evolutionary algorithm. In: Pelikan, M., Branke, J. (eds.) 12th Genetic and Evolutionary Computation Conference (GECCO 2010), Portland, OR, USA, pp. 233–240 (2010)
10. Wu, S., Banzhaf, W.: Rethinking Multilevel Selection in Genetic Program-ming. In: Conference of Genetic and Evolutionary Computation (GECCO 2011), Dublin, Ireland, July 12-16 (2011)
11. Gong, R., Zulkernine, M., Abolmaesumi, P.: A Software Implementation of a Genetic Algorithm Based Approach to Network Intrusion Detection. In: Sixth International Conference on Software Engineering, Artificial Intelligence,Networking and Parallel/Distributed Computing & 1st ACIS Int. Workshop on Self-Assembling Wireless Networks (2005)
12. Abdullah, B., Abd-alghafar, I., Salama, G., Abd-alhafez, A.: Performance Evaluation of a Genetic Algorithm Based Approach to Network Intrusion Detection System. In: 13th International Conference on Aerospace Sciences & Aviation Technology (ASAT 2013), Kobry Elkobbah, Cairo, Egypt, pp. 1–17 (2009)
13. Hoque, M., Mukit, M., Bikas, A.: An Implementation of Intrusion Detection System using Genetic Algorithm. International Journal of Network Security & Its Applications (IJNSA) 4(2), 109–120 (2012)
14. Bridges, S., Vaughn, R.: Fuzzy Data Mining And Genetic Algorithms Applied To Intrusion Detection. In: 12th Annual Canadian Information Technology Security Symposium, pp. 109–122 (2000)
15. Chittur, A.: Model Generation for an Intrusion Detection System Using Genetic Algorithms (2005),
http://ids.cs.columbia.edu/sites/default/files/
gaids-thesis01.pdf

16. Crosbie, M., Stafford, G.: Applying genetic programming to intrusion detection. In: AAAI Symposium on Genetic Programming, Cambridge, MA, pp. 1–8 (1995)
17. Dasgupta, D., Gonzalez, F.: An Intelligent Decision Support System for Intrusion Detection and Response. Information Assurance in Computer Networks 2052, 1–14 (2001)
18. Helmer, G., Wong, J., Honavar, V., Miller, L.: Automated discovery of concise predictive rules for intrusion detection. The Journal of Systems and Software (60), 165–175 (2002)
19. Moradi, M., Zulkernine, M.: A Neural Network Based System for Intrusion Detection and Classification of Attacks. In: 2004 IEEE International Conference on Advances in Intelligent Systems -Theory and Applications (2004)
20. Gomez, J., Dasgupta, D.: Evolving Fuzzy Classifiers for Intrusion Detection. In: The 2002 IEEE Workshop on Information Assurance (2002)
21. KDD Cup 1999 DataSet,
    http://kdd.ics.uci.edu/databases/kddcup99/kddcup99.html
22. Li, W.: A Genetic Algorithm Approach to Network Intrusion Detection. In: SANS Institute, USA (2003) (accessed)
23. Ashfaq, S., Farooq, M., Karim, A.: Efficient Rule Generation for Cost-Sensitive Misuse Detection Using Genetic Algorithms. In: International Conference on Computational Intelligence and Security, Guangzhou, vol. 1, pp. 282–285 (2006)
24. Pillai, M., Eloff, J., Venter, H.: An Approach to Implement a Network Intru-sion Detection System using Genetic Algorithms. In: SAICSIT 2004: the 2004 Annual research Conference of the South African Institute of Computer Scientists and and Information Technologists on IT Research in Developing Countries South African Institute for Computer Scientists and Information Technologists, Republic of South Africa, p. 221 (2004)
25. Goyal, A., Kumar, C.: GA-NIDS: A Genetic Algorithm based Network In-trusion Detection System (2008),
    http://www.cs.northwestern.edu/~ago210/ganids/GANIDS.pdf
26. Mischiatti, M., Neri, F.: Applying local search and genetic evolution in concept learning systems to detect intrusion in computer networks. In: Workshop About Machine Learning and Data Mining. Seventh Conference AI*IA Intelligenza Artificiale (2000)
27. Hu, Y., Chen, R., Tzeng, G.: Finding fuzzy classification rules using data mining techniques. Pattern Recognition Letters 24, 509–519 (2003)
28. Abadeh, M., Habibi, J., Lucas, C.: Intrusion Detection Using a Fuzzy Genet-ic-Based Learning Algorithm. Journal of Network and Computer Application 30, 414–428 (2005)
29. Gonzalez, F., Gomez, J., Kaniganti, M., Dasgupta, D.: An evolutionary approach to generate fuzzy anomaly signatures. In: 4th Annual IEEE Information Assurance Workshop, West Point, NY, pp. 251–259 (2003)
30. Hossain, M., Bridges, S.: A framework for an adaptive intrusion detection system with data mining. In: 13th Annual Canadian Information Technology Security Symposium (2001)
31. Abadeh, M., Mohamad, H., Habibi, J.: Design and analysis of genetic fuzzy systems for intrusion detection in computer networks. Expert Systems with Applications 38(6), 7067–7075 (2011)
32. Abadeh, M., Habibi, J.: Computer Intrusion Detection Using an Iterative Fuzzy Rule Learning Approach. In: IEEE International Fuzzy Systems Conference, FUZZ-IEEE 2007, London, pp. 233–240 (2007)

33. Tansel, O., Alhajja, R., Barker, K.: Intrusion detection by integrating boosting genetic fuzzy classifier and data mining criteria for rule pre-screening. Journal of Network and Computer Applications 30, 99–113 (2007)

34. Wang, L., Yu, G., Wang, G., Wang, D.: Method of evolutionary neural network based intrusion detection. In: International Conference of Information Technology and Information Networks, Beijing, China, vol. 5, pp. 13–18 (October 2001)

35. Marin, J., Ragsdale, D., Sirdu, J.: A hybrid approach to the profile creation and intrusion detection. In: DARPA Information Survivability Conference and Exposition II. DISCEX 2001, vol. 1 (2001)

36. Gomez, J., Dasgupta, D., Nasaroui, D., Gonzalez, F.: Complete expression Trees for evolving Fuzzy classifiers system with Genetic Algorithms and Applications to Network Intrusion Detection. In: NAFIPS-FLINT Joint Conference, New Orleans, LA, pp. 469–474 (2002)

37. Hofmann, A., Sick, B.: Evolutionary optimization of radial basis function networks for Intrusion Detection. In: Int. Joint. Conf. Neural Networks, Portland, OR, vol. 1, pp. 415–420 (2003)

38. Butz, M.: Rule-based Evolutionary online learning systems: learning bounds, classification, and prediction, PhD Thesis. In: University of Illinois at Urbana-Champaign (2004),
https://www.ideals.illinois.edu/bitstream/handle/2142/1087
6/Rule-based%20Evolutionary%20Online%20Learning%20Systems-
%20Learning%20Bound,%20Classification,%20and%20Prediction.
pdf?sequence=2

39. Ganapathy, S., Kulothungan, K., Yogesh, P., Kanna, A.: A novel weighted fuzzy C-means clustering based on immune genetic algorithm for intrusion detection. In: International Conference on Modeling Optimisation and Computing, pp. 1750–1757 (2012)

40. Amin, H., Radu, S.: On the optimality of cooperative intrusion detection for resource constrained wireless network. Computer & Security 34, 16–35 (2013)

41. Sen, S., Clark, J.: Evolutionary computation techniques for intrusion detec-tion in mobile ad hoc networks. Computer Networks 55(15), 3441–3457 (2011)

42. Balajinath, B., Raghavan, S.: Intrusion Detection Through Learning Behav-iour Model. International Journal of Computer Communications 24(12), 1202–1212 (2001)

43. Miller, B., Shaw, M.: Genetic Algorithms with Dynamic Niche Sharing for Multimodal Function Optimization. In: IEEE International Conference on Evolutionary Computation, Nagoya University, Japan, pp. 786–791 (1996)

44. Holland, J.: Adaptation in Natural and Artificial Systems: An introductory analysis with applications to biology, control, and artificial intelligence. Ann Arbor. Ann Arbor MI University of Michigan Press (1975)

45. Goldberg, D.: Genetic Algorithms in Search, Optimization, and Machine Learning. Addison-Wesley, Reading (1989)

46. Chih-Fong, T., Hsu, Y.-F., Lin, C.-Y., Lin, W.-Y.: Intrusion detection by machine learning: A review. Expert Systems with Applications 36, 11994–12000 (2009)

47. Freitas, A.: Data Mining and Knowledge Discovery with Evolutionary Algorithms. Springer, Heidelberg (2002)

Author Index